WJEC Mathematics for AS Level – Applied

Stephen Doyle

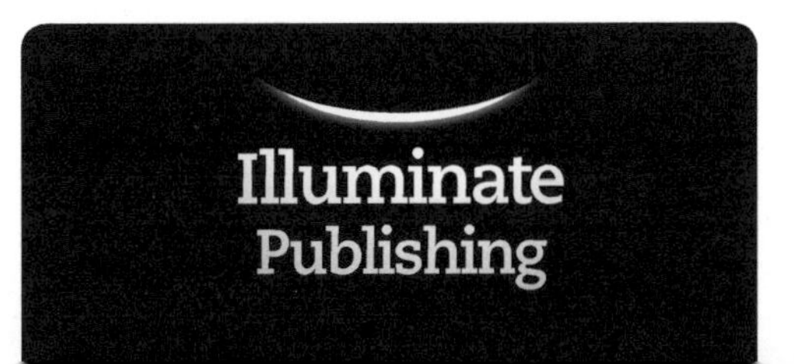

Published in 2018 by Illuminate Publishing Limited, an imprint of Hodder Education, an Hachette UK Company, Carmelite House, 50 Victoria Embankment, London EC4Y 0DZ

Orders: please contact Hachette UK Distribution, Hely Hutchinson Centre, Milton Road, Didcot, Oxfordshire, OX11 7HH. Telephone: +44 (0)1235 827827. Email: education@hachette.co.uk. Lines are open from 9 a.m. to 5 p.m., Monday to Friday. You can also order through our website: www.hoddereducation.co.uk

British Library Cataloguing in Publication Data

A catalogue record for this book is available from the British Library

ISBN 978 1 911208 52 5

Printed by Ashford Colour Press, UK

Impression 3
Year 2023

This material has been endorsed by WJEC and offers high quality support for the delivery of WJEC qualifications. While this material has been through a WJEC quality assurance process, all responsibility for the content remains with the publisher.

Hachette UK's policy is to use papers that are natural, renewable and recyclable products and made from wood grown in well-managed forests and other controlled sources. The logging and manufacturing processes are expected to conform to the environment regulations of the country of origin.

Editor: Geoff Tuttle
Cover design: Neil Sutton
Text design and layout: GreenGate Publishing Services, Tonbridge, Kent

Photo credits

Cover: Klavdiya Krinichnaya/Shutterstock; **p9** Marco Ossino/Shutterstock; **p10** DmitrySV/Shutterstock; **p16** Sergey Nivens/Shutterstock; **p55** Blackregis/Shutterstock; **p69** urickung/Shutterstock; **p91** Gajus/Shutterstock; **p113** cosma/Shutterstock; **p117** Will Rodrigues/Shutterstock; **p147** Andrea Danti/Shutterstock; **p169** sirtravelalot/Shutterstock.

Acknowledgements

The author and publisher wish to thank Sam Hartburn and Siok Barham for their help and careful attention in reviewing this book.

Contents

How to use this book

The contents of this study and revision guide are designed to guide you through to success in the Applied Mathematics component of the WJEC Mathematics for AS Level: Applied examination. It has been written by an experienced author and teacher. This book has been written specifically for the WJEC AS course you are taking and includes everything you need to know to perform well in your exams.

Knowledge and Understanding

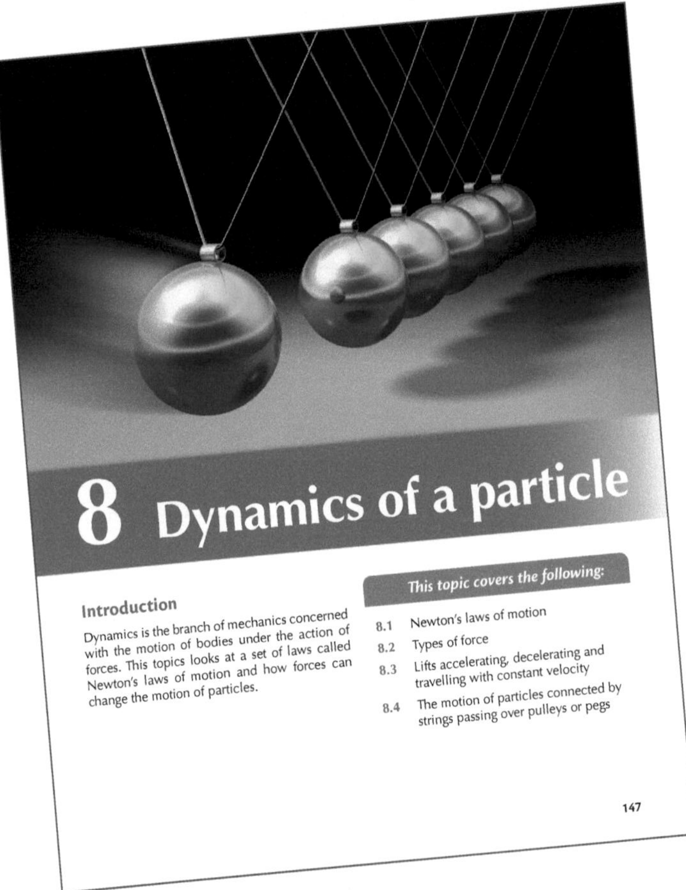

8 Dynamics of a particle

Introduction

Dynamics is the branch of mechanics concerned with the motion of bodies under the action of forces. This topics looks at a set of laws called Newton's laws of motion and how forces can change the motion of particles.

This topic covers the following:

8.1 Newton's laws of motion

8.2 Types of force

8.3 Lifts accelerating, decelerating and travelling with constant velocity

8.4 The motion of particles connected by strings passing over pulleys or pegs

147

Topics start with a short list of the material covered in the topic and each topic will give the underpinning knowledge and skills you need to perform well in your exams.

The knowledge section is kept fairly short leaving plenty of space for detailed explanation of examples. Pointers will be given to the theory, examples and questions that will help you understand the thinking behind the steps. You will also be given detailed advice when it is needed.

The following features are included in the knowledge and understanding sections:

- **Grade boosts** are tips to help you achieve your best grade by avoiding certain pitfalls which can let students down.

- **Step by steps** are included to help you answer questions that do not guide you bit by bit towards the final answer (called unstructured questions). In the past, you would be guided to the final answer by the question being structured. For example, there may have been parts (a), (b), (c) and (d). Now you can get questions which ask you to go to the answer to part (d) on your own. You have to work out for yourself the steps (a), (b) and (c) you would need to take to arrive at the final answer. The 'step by steps' help teach you to look carefully at the question to analyse what steps need to be completed in order to arrive at the answer.
- **Active learning** – are short tasks which you carry out on your own which aid understanding of a topic or help with revision.
- **Summaries** – are provided for each topic and present the formulae and the main points in a topic. They can be used for quick reference or help with your revision.

Exam Practice and Technique

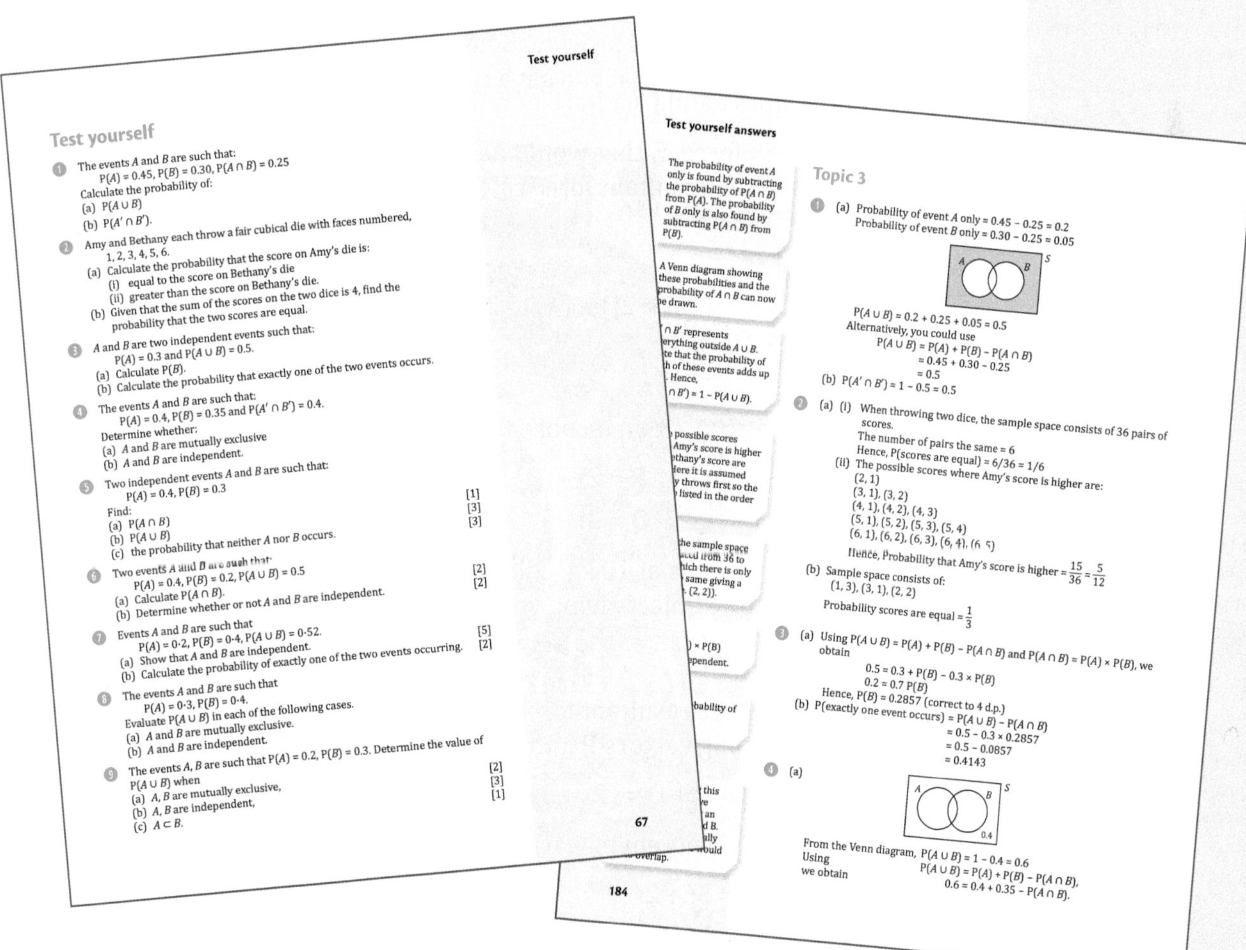

Test yourself

Test yourself

1. The events A and B are such that:
   $P(A) = 0.45, P(B) = 0.30, P(A \cap B) = 0.25$
   Calculate the probability of:
   (a) $P(A \cup B)$
   (b) $P(A' \cap B')$.
2. Amy and Bethany each throw a fair cubical die with faces numbered, 1, 2, 3, 4, 5, 6.
   (a) Calculate the probability that the score on Amy's die is:
   (i) equal to the score on Bethany's die
   (ii) greater than the score on Bethany's die.
   (b) Given that the sum of the scores on the two dice is 4, find the probability that the two scores are equal.
3. A and B are two independent events such that:
   $P(A) = 0.3$ and $P(A \cup B) = 0.5$.
   (a) Calculate $P(B)$.
   (b) Calculate the probability that exactly one of the two events occurs.
4. The events A and B are such that:
   $P(A) = 0.4, P(B) = 0.35$ and $P(A' \cap B') = 0.4$.
   Determine whether:
   (a) A and B are mutually exclusive
   (b) A and B are independent.
5. Two independent events A and B are such that:
   $P(A) = 0.4, P(B) = 0.3$
   Find:
   (a) $P(A \cap B)$ [1]
   (b) $P(A \cup B)$ [3]
   (c) the probability that neither A nor B occurs. [3]
6. Two events A and B are such that:
   $P(A) = 0.4, P(B) = 0.2, P(A \cup B) = 0.5$
   (a) Calculate $P(A \cap B)$. [2]
   (b) Determine whether or not A and B are independent. [2]
7. Events A and B are such that
   $P(A) = 0{\cdot}2, P(B) = 0{\cdot}4, P(A \cup B) = 0{\cdot}52$.
   (a) Show that A and B are independent. [5]
   (b) Calculate the probability of exactly one of the two events occurring. [2]
8. The events A and B are such that
   $P(A) = 0{\cdot}3, P(B) = 0{\cdot}4$.
   Evaluate $P(A \cup B)$ in each of the following cases.
   (a) A and B are mutually exclusive.
   (b) A and B are independent.
9. The events A, B are such that $P(A) = 0.2, P(B) = 0.3$. Determine the value of $P(A \cup B)$ when
   (a) A, B are mutually exclusive, [2]
   (b) A, B are independent, [3]
   (c) $A \subset B$. [1]

67

Test yourself answers

Topic 3

1. (a) Probability of event A only $= 0.45 - 0.25 = 0.2$
   Probability of event B only $= 0.30 - 0.25 = 0.05$
   $P(A \cup B) = 0.2 + 0.25 + 0.05 = 0.5$
   Alternatively, you could use
   $P(A \cup B) = P(A) + P(B) - P(A \cap B)$
   $= 0.45 + 0.30 - 0.25$
   $= 0.5$
   (b) $P(A' \cap B') = 1 - 0.5 = 0.5$
2. (a) (i) When throwing two dice, the sample space consists of 36 pairs of scores.
   The number of pairs the same = 6
   Hence, P(scores are equal) = 6/36 = 1/6
   (ii) The possible scores where Amy's score is higher are:
   (2, 1)
   (3, 1), (3, 2)
   (4, 1), (4, 2), (4, 3)
   (5, 1), (5, 2), (5, 3), (5, 4)
   (6, 1), (6, 2), (6, 3), (6, 4), (6, 5)
   Hence, Probability that Amy's score is higher $= \frac{15}{36} = \frac{5}{12}$
   (b) Sample space consists of:
   (1, 3), (3, 1), (2, 2)
   Probability scores are equal $= \frac{1}{3}$
3. (a) Using $P(A \cup B) = P(A) + P(B) - P(A \cap B)$ and $P(A \cap B) = P(A) \times P(B)$, we obtain
   $0.5 = 0.3 + P(B) - 0.3 \times P(B)$
   $0.2 = 0.7\ P(B)$
   Hence, $P(B) = 0.2857$ (correct to 4 d.p.)
   (b) P(exactly one event occurs) $= P(A \cup B) - P(A \cap B)$
   $= 0.5 - 0.3 \times 0.2857$
   $= 0.5 - 0.0857$
   $= 0.4143$
4. (a)
   From the Venn diagram, $P(A \cup B) = 1 - 0.4 = 0.6$
   Using $P(A \cup B) = P(A) + P(B) - P(A \cap B)$,
   we obtain $0.6 = 0.4 + 0.35 - P(A \cap B)$.

The probability of event A only is found by subtracting the probability of $P(A \cap B)$ from $P(A)$. The probability of B only is also found by subtracting $P(A \cap B)$ from $P(B)$.

A Venn diagram showing these probabilities and the probability of $A \cap B$ can now be drawn.

184

Helping you understand how to answer examination questions lies at the heart of this book. This means that we have included questions throughout the book that will build up your skills and knowledge until you are at a stage to answer full exam questions on your own. Examples are included; some of which are full examination style questions. These are annotated with Pointers and general advice about the knowledge, skills and techniques needed to answer them.

There is a Test yourself section where you are encouraged to answer questions on the topic and then compare your answers with the ones given at the back of the book. There are many examination-standard questions in each Test yourself that provide questions with commentary so you can see how the question should be answered.

You should, of course, work through complete examination papers as part of your revision process.

We advise that you look at the WJEC website www.wjec.co.uk where you can download materials such as the specification and past papers to help you with your studies. From this website you will be able to download the formula booklet that you will use in your examinations. You will also find specimen papers and mark schemes on the site.

WJEC Mathematics For AS Level Pure & Applied Practice Tests

There is another book which can be used alongside this book. This book provides extra testing on each topic and provides some exam style test papers for you to try. I would strongly recommend that you get a copy of this and use it alongside this book.

Good luck with your revision.

Stephen Doyle

1 Statistical sampling

Introduction

In order for statistics such as mean, mode, median, range, etc., to be obtained from data, the data must first be collected. If the data is not collected properly to start with then obtaining statistics from it will be a waste of time as they will not be accurate.

Obtaining accurate statistics depends on the proper collection of data and thought needs to be given as to how best to do this.

In this topic you will learn about how to take a smaller sample of the complete data set called the population so that accurate and meaningful statistics can be produced.

This topic covers the following:

- **1.1** The terms population and sample
- **1.2** Using samples to make informal inferences about the population
- **1.3** Sampling techniques: simple random sampling, systematic sampling and opportunity sampling
- **1.4** Selecting or critiquing sampling techniques

1.1 The terms population and sample

In order to prove or disprove a hypothesis or provide an answer to a question, data must be collected. It is usually too expensive and time consuming to measure or observe the whole set you are interested in (called the population). Instead a smaller set is used (called a sample) but it is essential that this sample is representative of the larger population.

Here are some important definitions:

Population – all members of the set that is being studied or has data collected about. So, for example, if you wanted to do a survey using all the students in your school then the population would be every student in your school. It is usually too costly and time consuming to survey the whole population, so a smaller set called a sample is used.

Sample – a smaller subset of the population that is used to draw conclusions about the population. As the sample is usually much smaller than the population, errors and inaccuracies can result when drawing conclusions about the population.

1.2 Using samples to make informal inferences about the population

Once a representative sample has been obtained, it can be used to make inferences about the population. Inferences are conclusions reached on the basis of evidence and reasoning. You do have to be careful when making inferences as the following example shows:

Example

1 The lake shown below contains a variety of fish. The owner of the lake would like to advertise the lake for fishing and wants to know about fish population of the lake.

Answer

1 The only way to find the exact information about fish in the lake would be to drain the lake and collect the fish so that the entire population of fish could be analysed – this is impractical. Instead a random sample of 300 fish was collected using nets and the results were:

Bream	46
Trout	128
Roach	126

Using this hopefully representative sample, we can make inferences about the fish population of the lake. Here are some inferences which could be made:

- There are more trout and roach compared to bream – this is a correct inference as there are significant differences in the numbers in the sample.
- The population of bream is about one third of the population of trout or roach – this is a reasonable inference. Here we are quantifying the inference (i.e. giving numeric information).
- There are more trout than roach in the lake – although this is true for the sample there is not a significant difference (only 2 fish). If we took another random sample we could get slightly different results resulting in more roach compared to trout so this is an inference we should not make.

It is important not to make an inference about the population based on a small difference in the sample. Remember that different samples will have slight differences in their make-up.

1.3 Sampling techniques: simple random sampling, systematic sampling and opportunity sampling

There are a number of different ways a sample may be taken:

- **Simple random sampling** – each member of the population has an equal probability of being included in the sample. Hence the members of the population to be included in the sample are picked 'randomly'. To generate a random sample the members of the population can be given numbers and then numbered balls can be picked out of a bag or a calculator can be used. There are also websites which you can use to generate random numbers in a specified range.
- **Systematic sampling** – is where sample members are selected from a larger population according to a random starting point and a fixed sampling interval. The sampling interval is found by dividing the population size by the desired sample size.

  For example, if there is a population of 200 households in a street and a sample size of 20 is taken then you can consider all the houses from 1 to 200 arranged in a circle. You could then choose to start at a randomly picked house (say number 12) and then calculate the sampling interval

  $$\left(\text{i.e. sampling interval} = \frac{\text{population}}{\text{sample size}} = \frac{200}{20} = 10\right)$$

  so you would then survey house 12 and then 22, then 32 and so on until you go right around in the circle back to house 12.

Active Learning

There are 200 houses in a street and you want to send a survey to 50 houses. You number each address from 1 to 200 and then select a random sample of 50 numbers. You do not want any duplicate numbers. Use the following random number website to randomly pick these 50 numbers without any duplicates.
www.random.org/integers
Evaluate this website explaining how easy, or not, it was to carry out this task.

Systematic sampling is most suited when the items being studied can be sorted into a sequence and where the allocation of a random number to each item would be difficult.

- **Opportunity sampling** – involves taking people from the population that are available at the time and are willing to take part in a survey. For example, you could ask people coming out of a station their opinion on the service offered by the train companies. This is a fast way of completing a survey but it can produce a biased sample, which means it does not fairly represent the population.

1.4 Selecting or critiquing sampling techniques

The ideal sample will:

- Be large enough
- Represent the population
- Be unbiased.

It is important to be able to look at a sampling technique critically to see if the sample is biased in any way.

Here are some things to consider:

The sample size – too small a sample may produce strange results, e.g. the sample could produce all people of a certain sex, age, income, etc.

The time of day the sample is taken – if you took a sample of train passengers at 8am you bias the sample towards people who were working.

Where the sample is taken – if you wanted information about the types of holidays people took, taking it at an airport would bias it towards those people taking their holidays abroad.

Advantages and disadvantages of different sampling techniques

Simple random sampling

Advantages

- Least biased of all sampling techniques – each member of the total population has an equal chance of being selected.
- Easily performed (e.g. picking numbers out of a hat, using a website, using the random number on a calculator or a random number function in a spreadsheet).
- Sample is highly representative of the population.

Disadvantages

- Time consuming and tedious to perform.
- Can lead to poor representation of the overall population, if certain members are not hit by the random numbers generated.

Systematic sampling

Advantages

- More straight-forward compared to simple random sampling.
- Sample is easy to select.

Disadvantages

- Less random than simple random sampling.

Opportunity sampling

Advantages

- Easy to take the sample as it is drawn from that part of the population that is close at hand.

Disadvantages

- Sample can be highly unrepresentative of the population as the sample is not picked at random.

Example

1 A researcher is collecting data about the amount of television watched by 20 different five year olds on a Saturday. The results in hours are listed below:

1, 10, 2, 5, 8, 0, 1, 1, 3, 2, 9, 9, 12, 5, 1, 2, 4, 9, 7, 12

(a) Taking an opportunity sample of the first 5 numbers in the list, calculate the mean number of hours watched on a Saturday.

(b) A systematic sample is to be taken of 5 data values.

(i) Work out the sampling interval.

(ii) A random number was chosen in the sampling interval and it was 3. Using this value write down the list of data values in the sample.

(iii) Using the list from (ii), work out the mean number of hours watched on a Saturday using this sample.

(c) State and give reasons which sampling method is likely to give more reliable results.

Answer

1 (a) Mean $= \dfrac{1 + 10 + 2 + 5 + 8}{5} = 5.2$

(b) (i) sampling interval $= \dfrac{\text{population}}{\text{sample size}} = \dfrac{20}{5} = 4$

(ii) 2, 1, 9, 1, 7

(iii) Mean $= \dfrac{2 + 1 + 9 + 1 + 7}{5} = 4$

(c) The systematic sample.

As it is a random sample and uses numbers throughout the distribution and not the first 5 values which may not be typical of the rest.

> The random number 3 means that you count along to the third value in the list. This is then the first data item in the sample. Now count along four numbers and this gives the second data value. This is repeated until the 5 data values are obtained.

Test yourself

1. Amy read in a local newspaper that the average number of hours per day of sunshine in her area in June is 8.
   She decides to test this by recording the number of hours of sunshine per day for the week beginning 1st June 2017.
   (a) Comment of the suitability of this sample to test the newspaper's claim.
   (b) How might Amy obtain more representative data in order to test the claim?

2. A popular club has a maximum capacity of 300 and is full to capacity every Saturday.
   A study of the people entering the club needs to be carried out and a sample size of 30 is to be used.
   (a) Describe how systematic sampling could be used to obtain the sample.
   (b) One person suggests that random sampling could be used with each person being given a number as they come into the club and then the 30 numbers are picked at random. Explain why this method of sampling would be difficult in this situation.

3. Julie thinks that everyone watches one or more soaps. She decides to test this by asking all her friends.
   (a) Give the name of the sampling method she has used.
   (b) Explain why her sample is likely to be biased.

4. A company is researching the education levels of its employees. Each employee is allocated a number and a random number generator is used to select people to interview.
   Give the name of this sampling technique.

5. The houses along a road called High Street are numbered from 1 to 350. The local council are going to undertake a survey about local amenities and they need a sample using 50 households.
   (a) Explain using the information above, the difference between the population and a sample.
   (b) Give two reasons why the council has decided to take a sample rather than use the population for the survey.
   (c) Explain how they might choose an unbiased sample.

6. A football ground has a maximum capacity of 40 000 supporters. A brief survey is to be conducted about the facilities at the ground. At the next match a sample of 50 supporters is to be used for the survey.
   (a) Explain what the population would be.
   (b) Give two reasons why a sample is used rather than the population for the survey.
   (c) The people who are in charge of conducting the survey want to complete the survey quickly so they decide to do the survey on the first 50 supporters entering the ground.
      (i) Give the name of this sampling technique.
      (ii) Explain why this sampling technique might not represent the population.

Summary

Check you know the following facts:

Population – all members of the set that is being studied or has data collected about it.

Sample – a smaller subset of the population and is used to draw conclusions about the population.

Sampling techniques

Simple random sampling – each item in the population is given a number and then the required number of items in the sample are picked at random using calculator, program, website.

Systematic sampling – the sampling interval is found by dividing the population by the sample size. You then pick a random number within this sample size and start from that as the first item in the sample. You then add the sampling interval to the random number to get the next number in the sample. This is repeated until you have the required sample.

Opportunity sampling – here you decide on the sample size and simply use the most convenient way of collecting the sample (e.g. friends, relatives, classmates, work colleagues, etc).

2 Data presentation and interpretation

Introduction

This topic builds on and reinforces your GCSE work on graphs and charts used in statistics. Rows, columns or tables of data are a starting point in statistics, but to make them a bit more exciting we can present the data in graphs and charts. Presenting them in this way not only improves their appearance, it also makes it easier to make comparisons of the data and spot trends.

As well as being able to construct these charts you must be able to select the most appropriate graph/chart to present a certain set of data. You must also be able to spot mistakes in graphs and charts produced by others.

This topic covers the following:

2.1 Interpreting diagrams for single variable data (histograms, box and whisker and cumulative frequency diagrams)

2.2 Scatter diagrams and regression lines

2.3 Making predictions using equations of regression lines

2.4 Measures of central tendency (mean, median and mode)

2.5 Measures of central variation (variance, standard deviation, range and interquartile range)

2.6 Selecting and critiquing data presentation techniques

2.7 Cleaning data (dealing with missing data, errors and outliers)

2.1 Interpreting diagrams for single variable data (histograms, box and whisker, and cumulative frequency diagrams)

Observations or measurements of a variable are called data.

Data can be of two types:

- **Quantitative** – where values (i.e. numbers) are given.
- **Qualitative** – where the data is non numeric (e.g. very hard, hard, easy, very easy).

Quantitative data can be further divided into discrete or continuous.

Discrete data – data that can only take certain values (e.g. the number of children in a household can be 0, 1, 2, 3, etc.) and in a certain range.

Continuous data – data that can take any value within a certain range.

Single variable data graphs and charts use a set of measurements (i.e. variable values) and the number of times they occur (i.e. the frequency). There are a number of graphs/charts that can be used to display single variable data and these include:

- Bar charts
- Histograms
- Pie charts
- Box and whisker diagrams
- Cumulative frequency curves.

Bar charts

Bar charts have qualitative data on one axis and frequency on the other. The following computer-produced bar chart shows the frequency on the vertical axis and the qualitative data (i.e. categories (pets in this case)) on the horizontal axis.

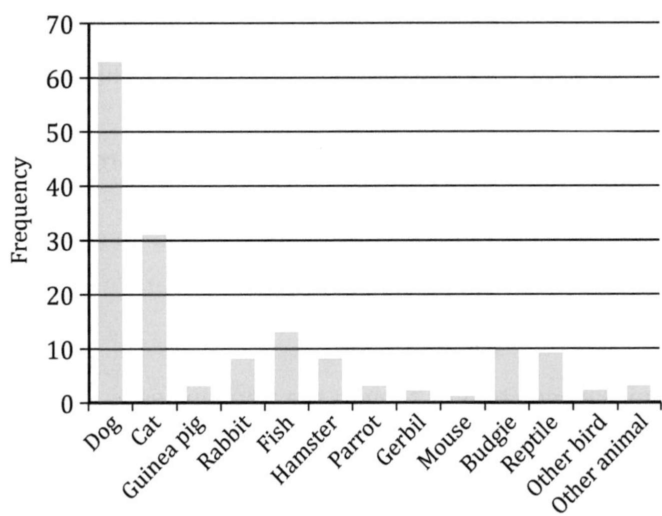

In the actual exam, you would not be expected to draw graphs using graph paper, i.e. accurate representations of the data in a table, but you may be asked to sketch a graph, showing, for example, the shape of a distribution, or perhaps drawing a box plot and marking on any key values.

Bar charts are useful to compare data for two different variables like this

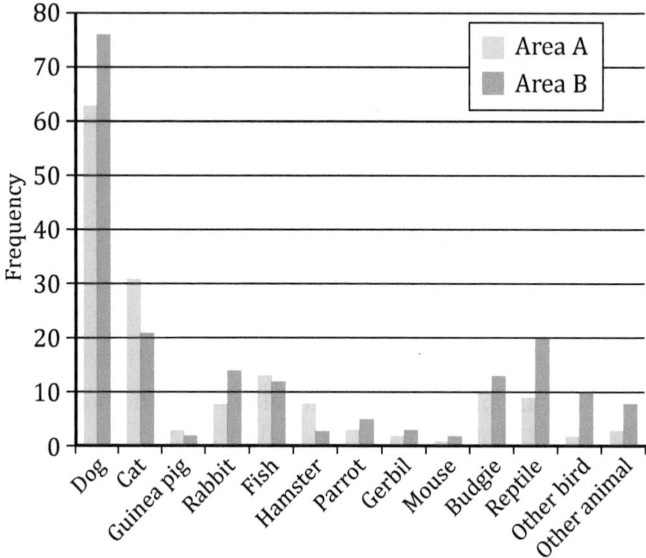

The main features of bar charts are:

- There are gaps between the bars for single variable bar charts.
- There are usually categories on one axis and frequency on the other.
- The height of the bar represents the frequency.

Histograms

On first impressions, histograms look like bar charts but there are some important differences:

- There are gaps between the bars (or groups of bars) in bar charts but in histograms there are no such gaps.
- There are numbers on both sets of axes – bar charts usually have categories on the horizontal axis.
- The bars are not usually equal widths. The bars in a bar chart are always equal widths.
- Frequency density is plotted on the vertical axis. Bar charts have frequency plotted on the vertical axis.
- Histograms are drawn when the data is continuous like an ordinary graph and the data can be put into classes (e.g. $100 \le h < 200$).
- The area of a bar represents the frequency.

Drawing a histogram

To draw a histogram, we need to first convert the frequencies to frequency densities because in a histogram it is the area of the bar rather than its height which represents the frequency.

Here is a bar of a histogram. The height is the frequency density and the width is the class width and the area of the bar represents the frequency.

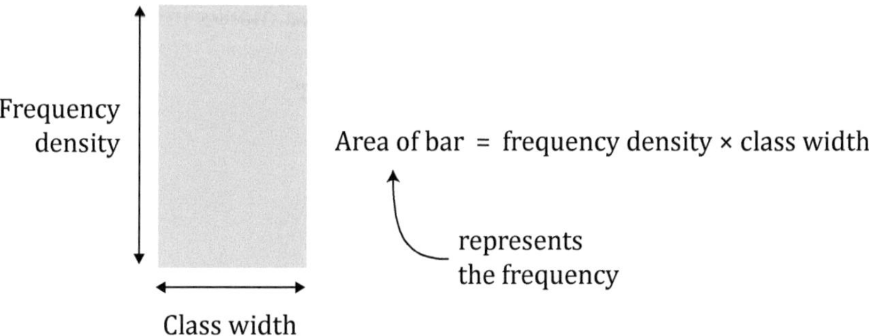

As frequency (area of bar) = frequency density × class width

To work out frequency density we use:

$$\text{Frequency density} = \frac{\text{frequency}}{\text{class width}}$$

The lengths of metal bars (l) in m and their frequencies were recorded in the following table:

Length (l m)	Class width (m)	Frequency	Frequency density
$1.0 \leq l < 1.2$	0.2	4	20
$1.2 \leq l < 1.4$	0.2	8	40
$1.4 \leq l < 1.8$	0.4	12	30
$1.8 \leq l < 2.4$	0.6	24	40
$2.4 \leq l < 2.6$	0.2	14	70
$2.6 \leq l < 3.6$	1.0	12	12

The data from the table can now be used to draw the following histogram:

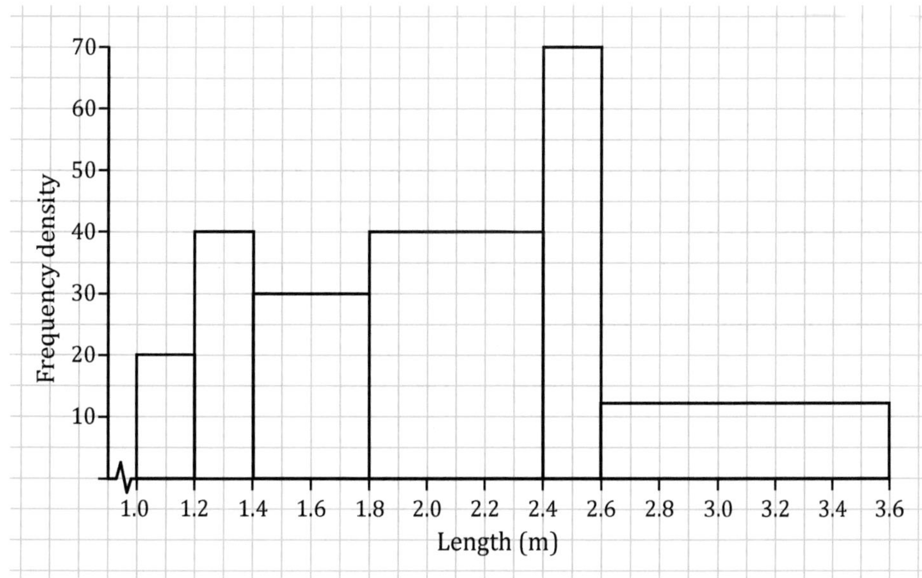

Example

1 Some students in a school class took a geography examination. The unfinished table and histogram show information about the marks in the examination.

Mark (%)	Frequency
$0 < x \le 20$	5
$20 < x \le 50$	
$50 < x \le 60$	25
$60 < x \le 80$	10
$80 < x \le 85$	
$85 < x \le 100$	30

Complete the table and the histogram.

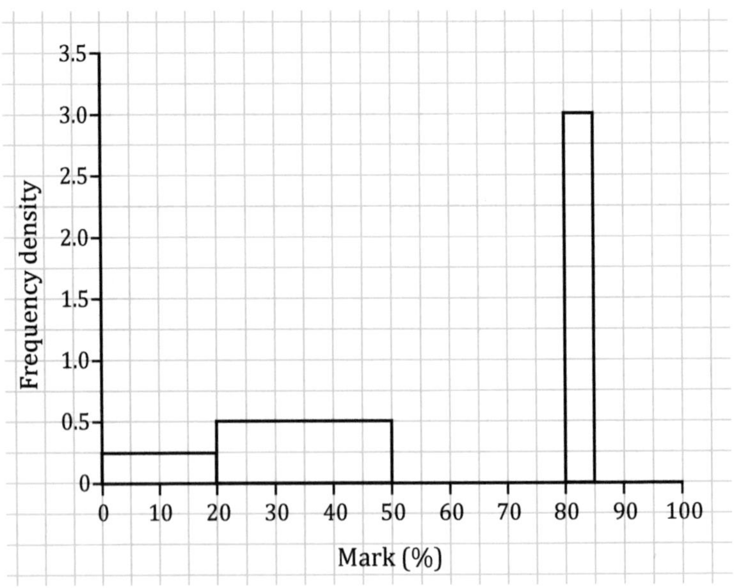

Answer

1 To find the frequency of the second bar,

frequency = area of bar = class width × frequency density = 30 × 0.5 = 15.

To find the frequency of the bar with class interval $80 < x \le 85$,

frequency = area of bar = class width × frequency = 5 × 3 = 15

Hence the completed table is

Mark (%)	Frequency
$0 < x \le 20$	5
$20 < x \le 50$	15
$50 < x \le 60$	25
$60 < x \le 80$	10
$80 < x \le 85$	15
$85 < x \le 100$	30

To complete the histogram, the frequency density needs to be calculated for the missing bars using the data in the table.

We use the formula $\quad \text{frequency density} = \dfrac{\text{frequency}}{\text{class width}}$

For class boundary $50 < x \leq 60$, $\quad \text{frequency density} = \dfrac{\text{frequency}}{\text{class width}} = \dfrac{25}{10} = 2.5$

For class boundary $60 < x \leq 80$, $\quad \text{frequency density} = \dfrac{\text{frequency}}{\text{class width}} = \dfrac{10}{20} = 0.5$

For class boundary $85 < x \leq 100$, $\quad \text{frequency density} = \dfrac{\text{frequency}}{\text{class width}} = \dfrac{30}{15} = 2$

The bars can now be added to the histogram.

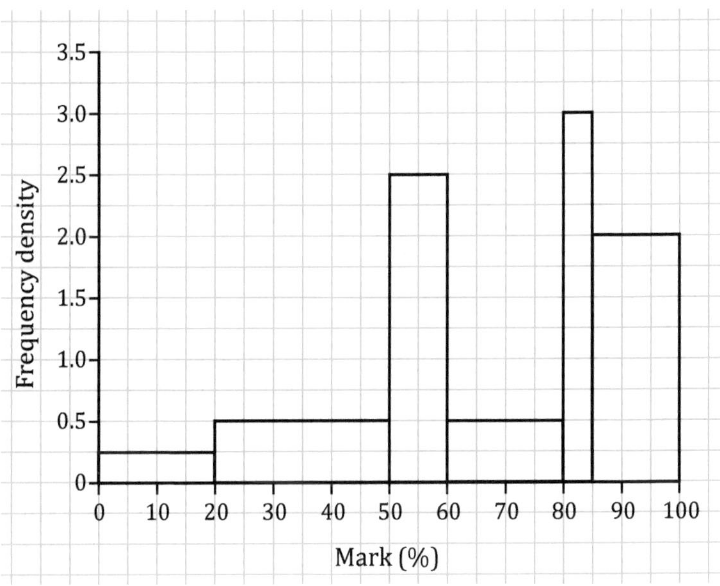

Box and whisker diagrams

A box and whisker diagram is a diagram that gives a visual representation to the distribution of data. The diagram highlights where most of the values lie and any values that greatly differ from the norm which are referred to as **outliers**.

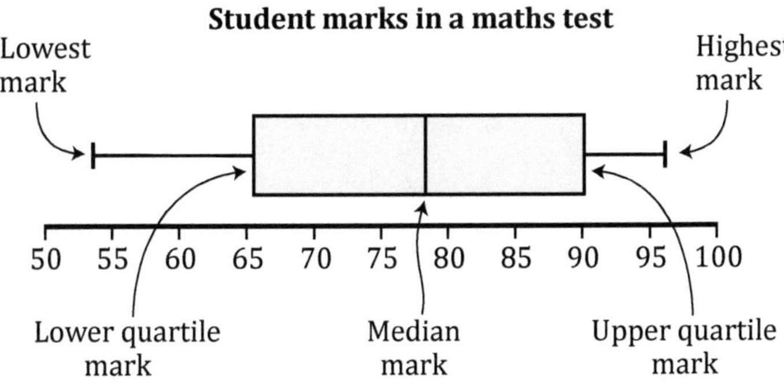

The left-hand side of the box represents the lower quartile and the right-hand side represents the upper quartile. The vertical line inside the box represents the median. The length of the side of the box represents the inter-quartile range.

The quartiles cut the data into four equal sections when the data is arranged in order of size. The first cut is the lower quartile (LQ or Q_1), the second cut is the median and the third cut is the upper quartile (UQ or Q_3). The difference between the upper and lower quartiles is called the interquartile range (IQR).

So IQR = UQ − LQ

or IQR = $Q_3 - Q_1$

The horizontal lines which protrude out from either side of the box extend to the minimum and the maximum values in the set of data as long as none of these values are outliers. These are the whiskers and their ends are marked by two shorter vertical lines.

Values higher than $Q_3 + 1.5 \times \text{IQR}$ or lower than $Q_1 - 1.5 \times \text{IQR}$ are called outliers and these are plotted to the right of the right whisker or the left of the left whisker.

Example

1 Plot a box and whisker diagram using the data below:

1, 5, 1, 2, 4, 3, 3, 2, 1, 5, 4, 5, 1, 6, 2, 0, 6

Answer

1 First order the data

0, 1, 1, 1, 1, 2, 2, 2, 3, 3, 4, 4, 5, 5, 5, 6, 6

Find n which is the number of data values in the list ($n = 17$ here)

The median is at the $\frac{n+1}{2}$ value.

In this case n is 17 so $\frac{17+1}{2} = 9$th value, which is 3.

Median = 3.

The lower quartile is at the $\frac{n+1}{4}$ value.

As n is 17 this is the 4.5th value which is the average of the 4th and 5th values.

Lower quartile = 1

The upper quartile is at the $\frac{n+1}{4} \times 3$ value.

As n is 17 this is the 13.5th value, which is the average of the 13th and 14th values.

Upper quartile = 5.

The interquartile range (IQR) = upper quartile − lower quartile = 5 − 1 = 4

The highest and lowest values are 6 and 0, respectively, so the range is 6 − 0 = 6

The box and whisker diagram can now be drawn:

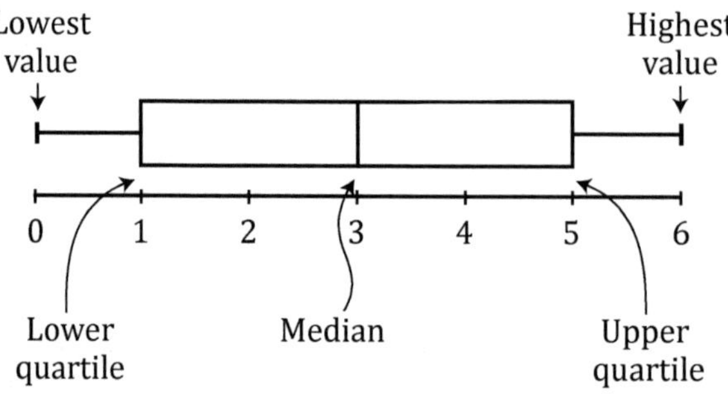

Using a computer to produce box and whisker diagrams

There are many different and free software packages that you can use to produce box and whisker diagrams. You usually enter the values as a list separated by commas. You can then tailor such things as headings, scales, etc., before the diagram is produced. The software usually produces a vertical box and whisker diagram and most of the packages will list the important measures such as lower quartile, range, median, etc., to save you having to read them off the diagram.

A box and whisker diagram is drawn using computer software based on the following set of data

1, 1.2, 1.5, 1.6, 1.9, 2.0, 2.1, 3.4, 4.5, 6.0, 9.2

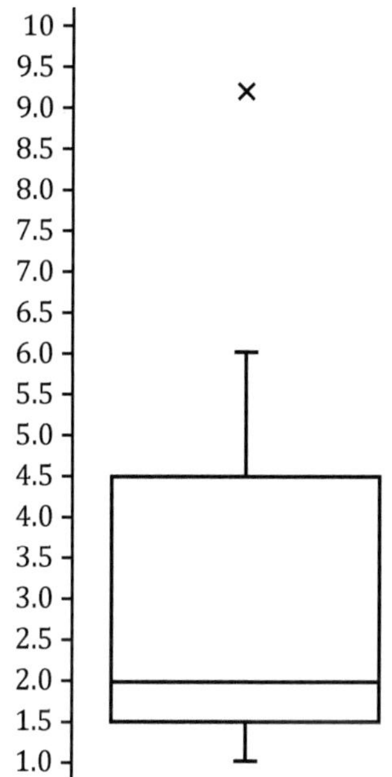

Notice there is a point at 9.2 that is outside the main diagram. This point has been identified as an unrepresentative value called an outlier and it is not included in the data used to plot the box plot. Luckily the software works out outliers and shows them as separate points on the diagram.

Look carefully at the diagram to check that you can obtain the following quantities from the diagram:

The outliers are ignored when drawing the rest of the box and whisker diagram. Notice the largest value in this set of data is now 6 as the outlier at 9.2 has been ignored.

Median = 2

Lower quartile = 1.5

Upper quartile = 4.5

Interquartile range = upper quartile − lower quartile = 4.5 − 1.5 = 3

Smallest value = 1

Largest value = 6

Range = largest value − smallest value = 6 − 1 = 5

Outlier = 9.2

Active Learning

Use the Internet to find a free software package that can draw a box and whisker diagram. Check that the software package is able to identify outliers and show them on the diagram.

Use the software package to produce a box and whisker diagram which shows the number of emails received per day over a two-week period.

12, 21, 32, 14, 56, 45, 11, 36, 10, 17, 76, 47, 50, 110

Produce a printout of the diagram.

Cumulative frequency diagrams

If you have a grouped frequency distribution and you are asked to find the median or the quartiles, then you can draw a cumulative frequency diagram and use that.

Cumulative frequency diagrams have a running total of the frequency (called the cumulative frequency) on the vertical axis and the quantity you want information about on the horizontal axis.

Suppose you have the following table showing the average distance in miles travelled on a full charge for a random sample of 200 electric cars. Notice the classes for the variable which here is the number of miles travelled (d).

Average distance travelled per charge (*d* miles)	**Frequency**
$0 \le d < 50$	3
$50 \le d < 100$	31
$100 \le d < 150$	60
$150 \le d < 200$	84
$200 \le d < 250$	22

We start by adding a running total of the frequency, called cumulative frequency to the table.

Average distance travelled per charge (*d* miles)	**Frequency**	**Cumulative frequency**
$0 \le d < 50$	3	3
$50 \le d < 100$	31	34
$100 \le d < 150$	60	94
$150 \le d < 200$	84	178
$200 \le d < 250$	22	200

We now plot the graph with 'average distance travelled per charge' on the x-axis and 'cumulative frequency' on the y-axis. We use the higher value of each class (or group) with its corresponding cumulative frequency.

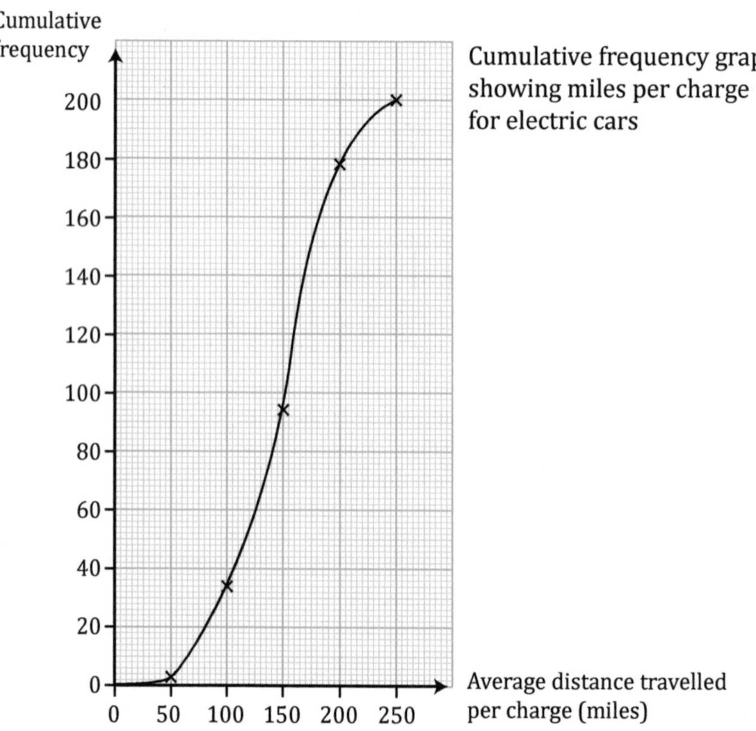

Cumulative frequency graph showing miles per charge for electric cars

The cumulative frequency curve can then be used to find the median, lower quartile, upper quartile and percentiles (e.g. the range of 70% of electric cars).

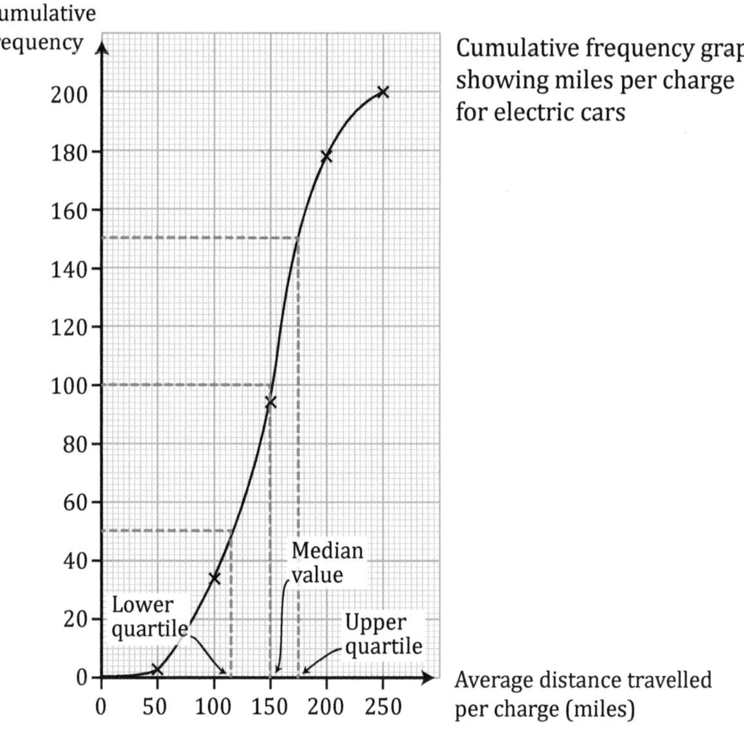

Cumulative frequency graph showing miles per charge for electric cars

Half way through the cumulative frequency (i.e. half of 200 = 100) we draw a horizontal line to meet the curve and then produce it down vertically to read off the median value. We then use a similar method to read off the values at one quarter and three quarters of the cumulative frequency (i.e. at 50 and 150 respectively) to give the lower and upper quartiles.

Symmetric, positive and negative skew

Graphs can be used to show probabilities. Here are the probabilities of obtaining certain values of the discrete variable *X*.

A graph of possible values for *X*, and the probabilities of obtaining them, is called a probability distribution.

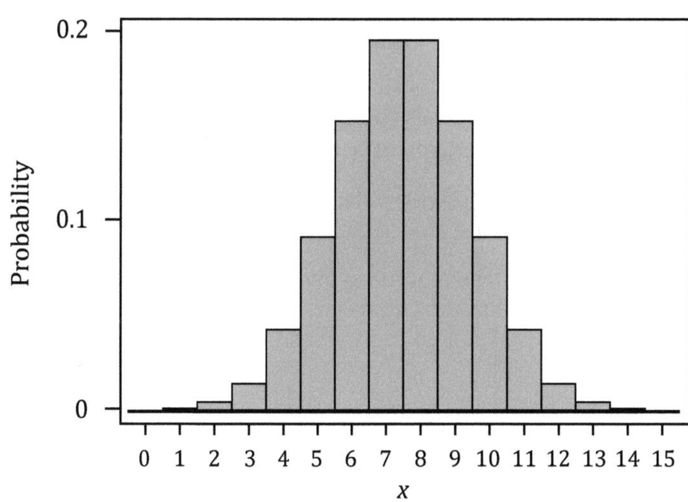

As the graph shows all the possible values *X* can take, the heights of all the bars will add up to one.

The probability distribution is symmetrical about the two most probable values 7 and 8. The distribution is said to be symmetric.

This distribution is positively skewed as it tails off in the direction of higher values for X.

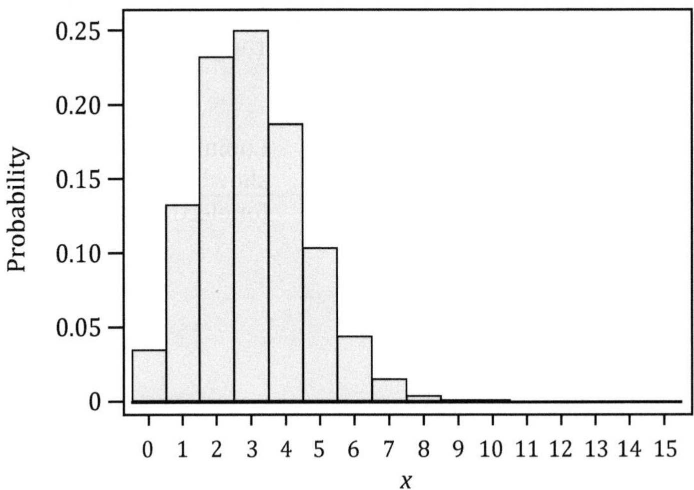

This graph is no longer symmetrical. It is positively skewed which means the bulk of the probability falls in the smaller numbers 0, 1, 2, ... , and the distribution tails off to the right.

This distribution is negatively skewed as it tails off in the direction of lower values for X.

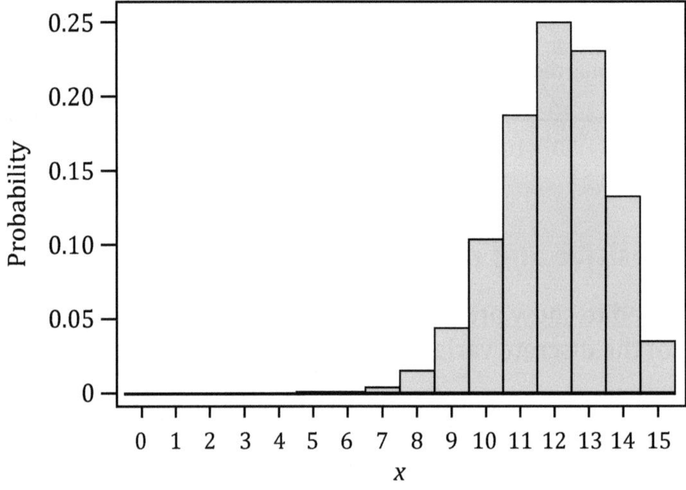

The above graph is negatively skewed as the bulk of the probability falls into the higher numbers and the probability tails off to the left.

Grade BOOST

To help remember which way graphs are skewed. You go by the direction of the tail – if it is towards larger values of X it is positively skewed and if it is towards smaller values of X it is negatively skewed.

Here are some graphs which show the overall shape of the various measures of skew.

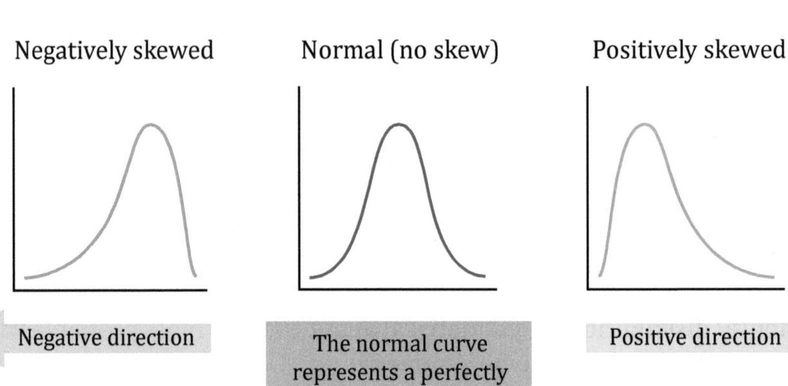

If the median is to the left of the mean, it means the distribution is positively skewed. If the median is to the right of the median it means the distribution is negatively skewed.

2.2 Scatter diagrams and regression lines

A scatter diagram is a visual representation of any relationship between two variables. Scatter diagrams are used to show possible relationships between bivariate data. Bivariate data is data that has pairs of values for two different variables. For example, you could collect pairs of values for shoe size and height to see if there is correlation between the two variables.

Interpretation of correlation (positive, negative, zero, strong and weak)

Correlation is a measure of how well two variables are related to each other. If when one variable increases, the other variable also increases this is called positive correlation. An example of this would be height and foot size – taller people generally take a greater shoe size.

If an increase in one variable is associated with a decrease in the other this is called negative correlation. An example of this would be distance from the Equator and temperature. Generally, the further from the Equator, the lower the temperature.

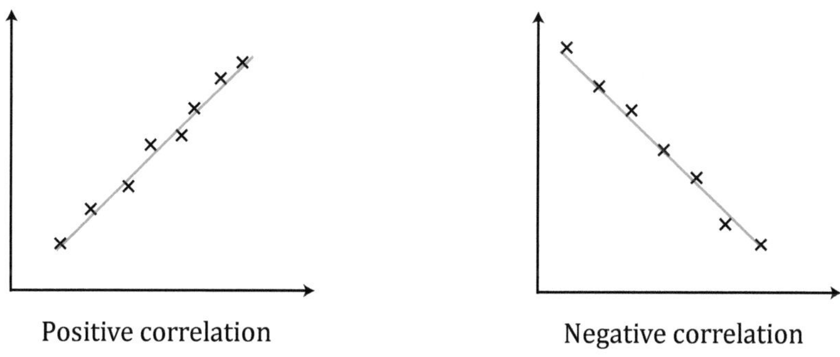

Positive correlation

Negative correlation

If there is no relationship between the variables this is called zero correlation. For example there is no relationship between the height of a person and the number of pets owned, so this is zero correlation.

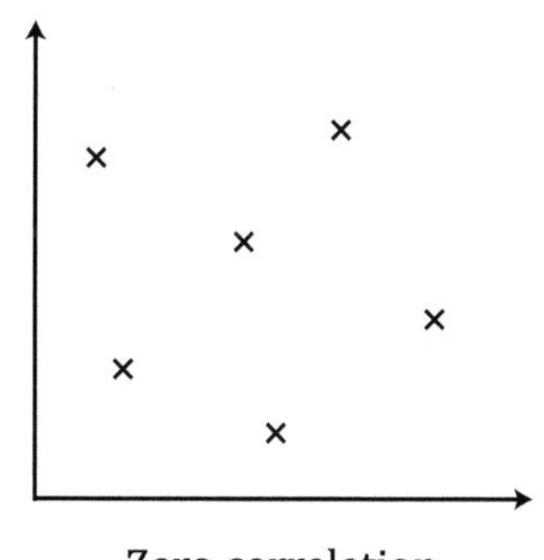

Zero correlation

Strong and weak correlation

Strong correlation is shown by points that lie close to a straight line.

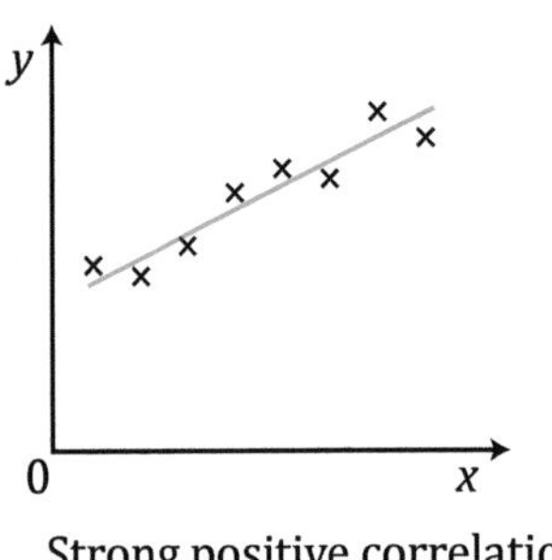

Strong positive correlation

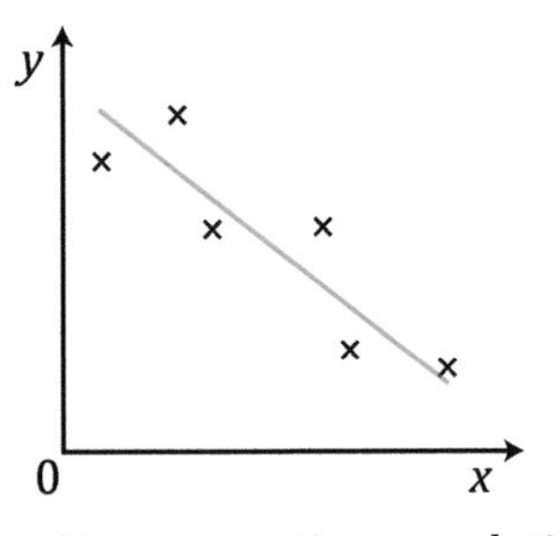

Strong negative correlation

Weak correlation is shown by points that do not lie close to a straight line.

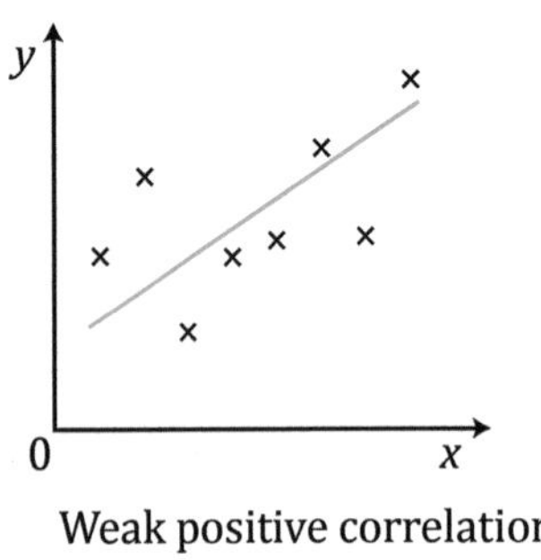

Weak positive correlation

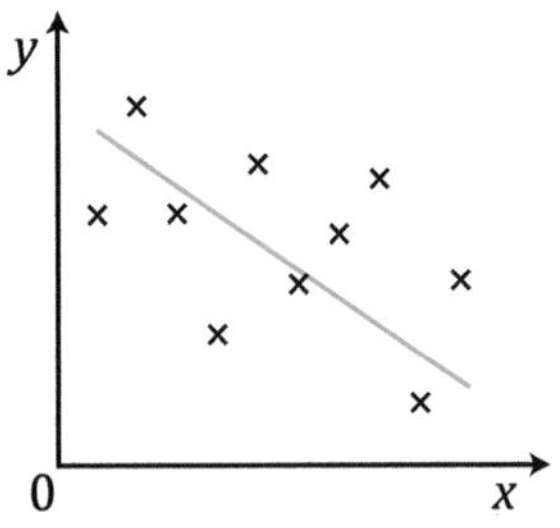

Weak negative correlation

Regression lines

When two quantities are plotted against each other to produce a scatter graph and it is clear that there is a linear relationship between the variables then a line of best fit can be drawn. The straight line models the relationship between the two variables.

If the two variables are x and y, then the line of best fit will have an equation of the form

$$y = mx + c$$

where m is the gradient of the line and c is where the line cuts the y-axis.

Once the equation of the line has been found it can be used to find a value of y that corresponds to a certain value of x and vice versa.

Example

1 The table shows the temperature (x) and the number of visitors (y) at a beach.

Temperature (x) °C	5	10	15	20	25	30	35
Visitor numbers (y)	17	30	40	53	60	85	90

(a) Present the data shown in the table as a scatter diagram.

(b) Draw in the line of best fit.

Answer

1 (a) and (b)

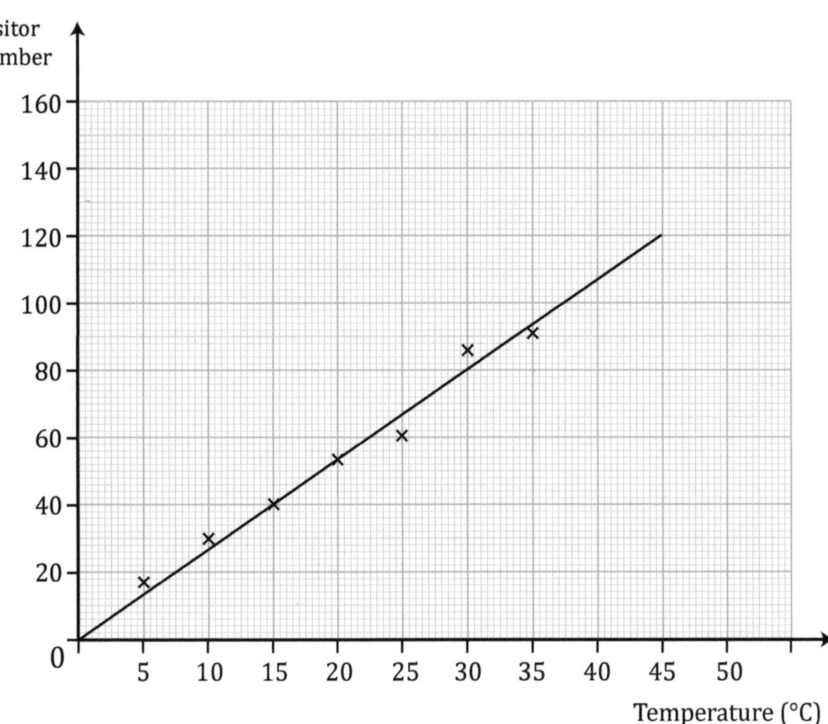

Using the regression equation

The regression equation can be used to calculate one value of a variable given the value of the other variable.

However, the regression equation must be used with caution.

We only know that the values lie on an approximate straight line between the range of the data values. If we extrapolate (i.e. obtain values outside the range of the data) we are making the assumption that the regression line is true for all the values, which may not be the case.

Take the example of people on the beach – we could use the regression equation to calculate the number of people on the beach when the temperature rose to 100 °C (which is impossible). The line passes through (0, 0) and (45, 120) which gives a gradient of 2.78. The regression equation for the line above is $y = 2.78x$.

So if $x = 100$, $y = 2.78 \times 100 = 278$ when in fact no-one would be at the beach, as at those temperatures they would all be dead!

2.3 Making predictions using equations of regression lines

You can use the correlation between two variables to make a prediction using the regression line. Prediction makes an estimate of the value of one variable based on the value of another variable. The stronger the correlation between two variables, then the closer the points lie to the regression line, and the more accurate the prediction will be.

There are two important terms used:

Interpolation is where a value is estimated using the line within the range of the data. So in our beach example you could find the number of people on the beach when the temperature is 22 °C because it lies within the existing data values 5 to 35 °C.

Extrapolation is where a value is estimated using the line outside of the range of the data. For example, using the line to find the number of people on the beach when the temperature is −5 °C. The **more you extrapolate** the **less reliable the result** is as the data may follow a different trend outside of the range collected

You can extrapolate or interpolate **using the equation for regression line** and substituting in either an x or y value to find the other value.

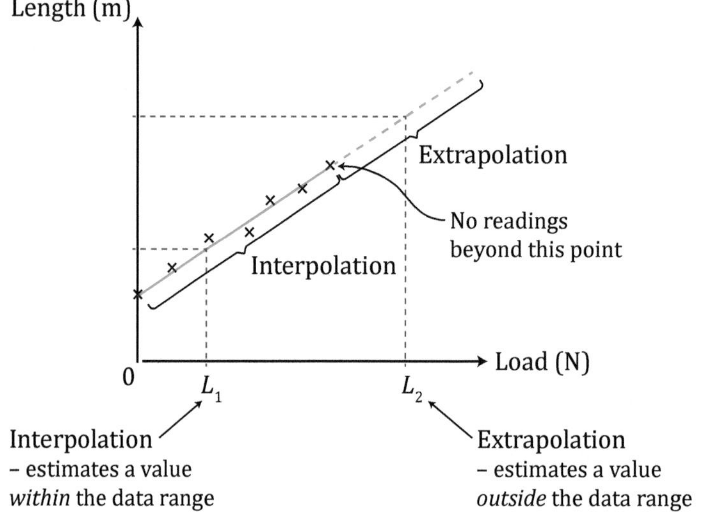

Correlation does not imply causation

If a change in one quantity is accompanied by a change in another quantity they are said to have correlation. However, just because there is correlation between two quantities it does not necessarily mean one quantity causes the other. Linking one thing with another does not always prove that the result has been caused by the other.

For example, there is a correlation between the sale of ice cream in America and the number of serious crimes, but you would not say that eating ice cream causes you to commit serious crime. Instead there is another factor – as the temperature goes up people are more irritable and more likely to commit crime.

You always have to use common sense deciding whether correlated quantities are causal or not.

Examples

1 The scatter diagram below shows data from a random sample of towns in Wales.

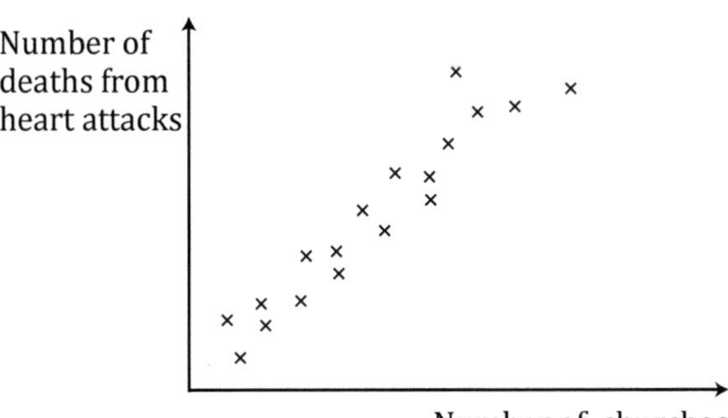

(a) Describe the correlation between the number of deaths by heart attack and the number of churches.

(b) State whether the relationship shown by the scatter diagram is causal or non-causal. Explain how you decided.

Answer

1 (a) Strong positive correlation.

(b) Non-causal as number of churches in a town would normally depend on the population of the town and the greater the population of the town, the greater the number of deaths by heart attack.

If you decreased the number of churches you would not decrease the number of deaths by heart attack.

---

2 A researcher found and downloaded a dataset for small sample of illnesses suffered in the UK in 2012. This dataset included the variables: 'Name of illness', 'Annual deaths' and 'Name length', the latter being the number of letters in the name of the illness.

(a) The scatter graph represents 'Annual deaths' against 'Name length'.

(i) Comment on the correlation between 'Annual deaths' and 'Name length'. [1]

(ii) Interpret the correlation between 'Annual deaths' and 'Name length' in this context. [1]

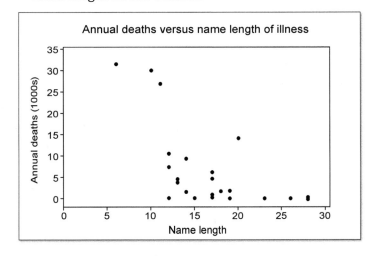

Annual deaths versus name length of illness

Annual deaths (1000s)

0 5 10 15 20 25 30 35

0 5 10 15 20 25 30

Name length

(b) The regression equation for this dataset is:

'Annual deaths' = 23.95 – 1.03 × 'Name length'

(i) Interpret the gradient of the regression equation for this model. [1]

(ii) State with a reason, whether the regression model would be useful to predict the annual number of deaths for a new disease. [1]

(iii) State whether the relationship between the 'Annual deaths' and 'Name length' is causal. Explain your answer. [1]

Answer

2 (a) (i) Weak negative correlation.

(ii) The fewer the number of letters in the name of the illness the higher the number of annual deaths.

(b) (i) Each additional letter corresponds to a decrease in the number of deaths by 1030 on average.

(ii) The model could not be used to predict annual deaths as the correlation is too weak. If you look at the annual deaths near zero, there is a very large difference in the corresponding name lengths.

(iii) It is not causal, as the length of the name of an illness has no relation to how serious the illness is, nor to how many people are likely to suffer from it.

2.4 Measures of central tendency (i.e. mean, median and mode)

A single value in a list of values that describes the centre of the data is called a **measure of central tendency**. There are three measures of central tendency:

- Mean
- Median
- Mode

Calculating the mean

There are two types of mean:

The **population mean, μ** and the **sample mean, $\bar{x}$.**

If the data used to calculate the mean is the entire population (e.g. every person in a company to work out the mean wage or every person in the school to work out the mean height), then the mean is called the population mean and it is given the symbol μ.

If the population is too large to deal with, a representative sample can be taken that should represent the population and this sample is used to calculate the mean. When a sample is used the mean is called the sample mean and is given the symbol $\bar{x}$.

If all the individual values (i.e. x) in the entire population are known then the population mean can be calculated using the formula

$$\mu = \frac{\sum x}{n}$$

where μ is the population mean, $\sum x$ is the sum of all the individual values and n is the number of values.

If the individual values (i.e. x) in a sample are known the sample mean can be calculated using the formula

$$\bar{x} = \frac{\sum x}{n}$$

where $\bar{x}$ is the sample mean, $\sum x$ is the sum of all the individual values and n is the number of values.

The mean of a frequency distribution where the data set is given in a frequency table is calculated using the formula:

$$\bar{x} = \frac{\sum fx}{\sum f} \quad \text{or} \quad \mu = \frac{\sum fx}{\sum f}$$

where $\bar{x}$ is the sample mean or μ is the population mean, $\sum fx$ is the sum of all the x values, each multiplied by its frequency, f, and $\sum f$ is the sum of all the frequencies.

This formula will not be given so it must be remembered.

If the entire population is used to calculate the mean we give the mean the letter μ. If a sample of the population is used to calculate the mean the mean is given the letter $\bar{x}$.

Notice both formulas have the same part after the equals. The different letters/symbols used just tell you whether the data used is the population or a sample.

These formulas will not be given, so they need to be remembered.

Example

1 Find the mean number of peas per pod from the table which shows the number of peas in a pod for a certain type of pea plant.

Notice that this is a frequency distribution. There is a single variable (peas in a pod) and the frequency (number of pods).

Number of peas in a pod	Number of pods
5	3
6	17
7	12
8	10
9	8
10	4

Answer

1 First decide which column of data represents the data values x and which column of data represents the frequency f.

As we are interested in the mean number of peas in a pod, the number of peas in a pod is the x-value and the number of pods is the frequency, f.

$$\bar{x} = \frac{\sum fx}{\sum f}$$

$$= \frac{(3 \times 5) + (17 \times 6) + (12 \times 7) + (10 \times 8) + (8 \times 9) + (4 \times 10)}{3 + 17 + 12 + 10 + 8 + 4}$$

$$= \frac{393}{54}$$

$$= 7.3 \text{ (1 d.p.)}$$

Working out the mode

The mode is the value(s) or class that occurs most often.

The mode of 1, 1, 2, 9, 12, 1, 5 is 1 as it occurs the most often (i.e. three times).

The modes of 1, 1, 2, 9, 12, 2, 5, 7 are 1 and 2 as both of these values occur the most often (i.e. twice). Note that you can have more than one mode so data can be multimodal.

In some cases the mode of a set of data is not useful. For example, the mode of 1, 2, 3, 4, 5, 6 has no use because each value could be considered the mode.

Modes are frequently used if there are one or two values that occur most often.

Working out the median

The median is the middle value when the data values are put in order of size.

If there are an odd number of values, there will be a middle value and this will be the median.

(e.g. for the data set 2, 4, 4, 5, 9 the median is 4)

If there is an even number of values, there will be two values in the middle so the mean of these two values is the median.

(e.g. for the data set 2, 4, 4, 5, 7, 9 the mean of the two middle values (i.e. 4 and 5) is found. This gives a median of 4.5.)

Examples

1 Find the median of 12, 10, 1, 4, 10, 8, 3, 5

Answer

1 The data set in order of size is 1, 3, 4, 5, 8, 10, 10, 12

The two middle values are 5 and 8 so the median is the mean of these two values $\left(\text{i.e. } \frac{5+8}{2} = 6.5\right)$.

You can also use the following method:

Find n which is the number of data values in the list ($n = 8$ here)

The median is at the $\frac{n+1}{2}$ value.

In this case n is 8 so $\frac{8+1}{2} = 4.5$th value which is the mean of the fourth and fifth values (i.e. 5 and 8) so the median is $\frac{5+8}{2} = 6.5$

2 The shoe sizes of a certain style of shoe in shop are as follows:

6, 6, 7, 8, 8, 8, 9, 9, 9, 9, 10, 10, 11, 12

Find:

(a) the mean size

(b) the modal size

(c) the median size.

Answer

2 (a) $\mu = \frac{\sum x}{n} = \frac{6+6+7+8+8+8+9+9+9+9+10+10+11+12}{14}$

$= \frac{122}{14}$

Mean = 8.7 (1 d.p.)

(b) The most frequent size as it appears four times.

Modal size = 9

(c) The data is already ordered and as there are an even number of values, there are two middle values which are 9 and 9.

Median size = 9

3 A farmer records the number of eggs laid each day by his chickens over a 31-day period and produces the following table:

Number of eggs (x)	12	13	14	15	16	17	18
Frequency (f)	3	7	8	10	2	1	0

Find:

(a) the mean

(b) the median

(c) the mode.

Answer

3 (a) $\mu = \frac{\sum fx}{\sum f}$

$= \frac{(3 \times 12) + (7 \times 13) + (8 \times 14) + (10 \times 15) + (2 \times 16) + (1 \times 17) + (0 \times 8)}{3+7+8+10+2+1+0}$

= 14.1 (1 d.p.)

(b) The data is arranged in order of size. Remember that it is the x-values that are ordered and as the total frequency is 31 (i.e. an odd number) there will be a single number in the middle. To work out the middle value you can think of the 31 values arranged as 15 1 15 so the 16th value is the middle value (i.e. the median). The 16th value will be 14.

Median = 14

(c) The most frequent number of eggs = 15

Mode = 15

You have to think of the 31 items of data arranged like this

12, 12, 12, 13, 13, 13, 13, 13, 13, 13, 14, ...

You now look for the 16th value.

Comparison of the measures of central tendency (mean, mode and median)

The three measures of central tendency (mean, mode and median) are often simply referred to as averages, so if you are told an average value you should always ask which average has been used. As there is a choice of three so the person stating the average value will choose the one that best suits their purpose.

There are advantages and disadvantages in using the mean, mode and median and here they are summarised in this table.

Mean	Mode	Median
Advantages	***Advantages***	***Advantages***
All the data values are used to calculate it – so all the data are taken into account	It can be used with data that are not numeric (e.g. favourite crisp flavour)	It is not affected by extreme values as only the middle value (or an average of two middle values) in an ordered list is chosen. It is a good one to choose if there are outliers.
Disadvantages	***Disadvantages***	***Disadvantages***
Is affected by extreme values	There can be many different modes, so it may not be any use	Can take a long time to order the data and work out what the median is
It is only useful with numeric data	There may be no mode	
	The mode may not represent the rest of the data well	

Example

1 The table below shows the annual salary details for employees working in a large dental practice.

Annual salary (£)	Number of people
0–9 999	7
10 000–19 999	6
20 000–29 999	6
30 000–39 999	0
40 000–59 999	2
60 000–80 000	3
80 000–160 000	5

(a) Write down the median class and the modal class.

(b) What is the disadvantage in using this modal class as an average to represent this set of data?

(c) An estimate of the mean can be found. Explain why it is only an estimate and explain why the mean annual salary might not represent this set of data well.

Answer

1 (a) As there are 29 people and the data is ordered it will be the class for the 15th person.

Median class = £20 000–£29 999

Modal class = £0–£9999

(b) The modal class has 7 people but two other classes are fairly close at 6 people which mean that the modal class does not represent the data well.

(c) The data forms a grouped distribution so the exact values of the salaries of each person are unknown. The mid value of each class is used to calculate the mean and this means that the mean is only an estimate.

There are some extreme values (both low and high) and this could result in a mean that is unrepresentative of the data.

2.5 Measures of central variation (variance, standard deviation, range and interquartile range)

Measures of central variation measure the spread of the data.

Two simple measures of spread are the range and the interquartile range.

Range – this is the difference between the largest value and the smallest value in a set of data.

Interquartile range (IQR) – is the difference between the upper quartile (Q_3) and the lower quartile (Q_1).

Hence $$\text{IQR} = Q_3 - Q_1$$

The interquartile range is a useful measure of spread as it represents the spread of the middle half of the data (i.e. one quarter either side of the median).

The interquartile range can be found using a cumulative frequency graph which has a running total for the frequency (called the cumulative frequency) plotted along the y-axis and the data values plotted along the x-axis.

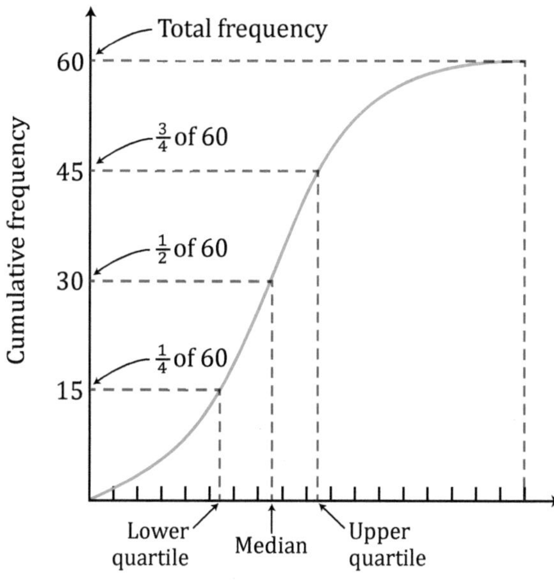

The upper quartile (Q_3) is found by finding $\frac{3}{4}$ of the total frequency and then using the graph to find the corresponding value on the x-axis. To find the lower quartile (Q_1), $\frac{1}{4}$ of the total frequency is found and then using the graph the corresponding x-value is found.

Then we use the formula IQR = $Q_3 - Q_1$ to find the interquartile range.

There are another two measures of central variation that are a bit more complicated to find.

Variance and standard deviation

It is useful to have a statistic which can measure the scatter of data and be small when the scatter is small (i.e. the data is clustered together) and large when the data is widely scattered.

The variance is a statistic that measures how far each value in the set of data is from the mean.

To work out the variance for a set of data take the following steps:

1. Subtract the mean from each value in the set of data. For example, if the first data value is X_1 and the mean is μ then we would have $X_1 - \mu$. We can write this as $X_i - \mu$. Note that X_i is a particular value for X in the sample so if there were three values of X we would have X_1, X_2 and X_3.
2. Square each of these differences and add all the squares together.

   This can be written as $\sum(X_i - \mu)^2$.
3. Divide the sum of the squares by the number of values in the data set (i.e. n)

These steps can be represented by the following formula:

$$\sigma^2 = \frac{\sum(X_i - \mu)^2}{n} \qquad \text{where } \sigma^2 \text{ is the variance.}$$

It is important to note that variance has the units of X^2 which means whatever units the data values X has, the variance will have the units squared. So say X has metre as the unit, then the variance will have the units of metre^2.

In order to have a measure which uses the same units as the data values, another quantity, **standard deviation**, is used. Standard deviation is the square root of the variance and will therefore have the same units as the units used for X.

Standard deviation

The standard deviation is a measure of the spread of values in a set of data. In most cases we are interested in the standard deviation of the population. However, collecting data values from the whole population may be prohibitive in terms of cost and time, so a smaller sample of data is used instead to find an estimate of the standard deviation.

If a sample is used to estimate the standard deviation for the whole population it is called the sample standard deviation. So there are two standard deviations; one based on the entire population which will be the accurate one (called σ) as it uses all the values and an estimate of the standard deviation using only values in a smaller sample (called S). This is called the sample standard deviation. In this topic we will only be dealing with the standard deviation, based on the population.

It is important to note that the standard deviation like the mean should only be used if the data is not significantly skewed or has outliers.

The square root of the variance (i.e. $\sqrt{\sigma^2} = \sigma$) is the standard deviation.

Hence the standard deviation $\sigma = \sqrt{\dfrac{\sum(X_i - \mu)^2}{n}}$

Where $\sum(X_i - \mu)^2$ is the sum of the squares of the differences between each value in the set and the mean and n is the total number of values.

This formula will not be given so you will need to remember that to obtain the standard deviation σ you need to square root the variance σ^2.

Example

1 The shoe sizes of 10 people in class are as follows:

6, 7, 8, 8, 9, 10, 11, 12, 8, 9

Find the standard deviation.

Notice that these 10 shoe sizes are the entire population. This means we need to find the standard deviation of the population σ.

Answer

1 As the mean is not given, it must be calculated

$$\mu = \frac{\sum x}{n} = \frac{6+7+8+8+9+10+11+12+8+9}{10} = \frac{88}{10} = 8.8$$

X	$X - \mu$	$(X - \mu)^2$
6	−2.8	7.84
7	−1.8	3.24
8	−0.8	0.64
8	−0.8	0.64
9	0.2	0.04
10	1.2	1.44
11	2.2	4.84
12	3.2	10.24
8	−0.8	0.64
9	0.2	0.04
		$\sum(X - \mu)^2 = 29.6$

Add up the values in this column to give $\sum(X - \mu)^2$

Standard deviation, $\sigma = \sqrt{\dfrac{\sum(x_i - \mu)^2}{n}} = \sqrt{\dfrac{29.6}{10}} = 1.72$

$n = 10$ as there are 10 data values in the sample.

This calculation is a bit tedious to perform and luckily the formula can be simplified to make the calculation easier.

The simplified formula for variance is:

$$\text{variance} = \frac{\sum x_i^2}{n} - \left(\frac{\sum x_i}{n}\right)^2$$

Where $\dfrac{\sum x_i^2}{n}$ is the mean of the squares of the values and $\left(\dfrac{\sum x_i}{n}\right)^2$ is the square of the mean of the values.

So, we can say the variance is the mean of the squares minus the square of the mean.

Using the previous example, but this time using the simplified formula, we can set up the table like this:

x	x_i^2
6	36
7	49
8	64
8	64
9	81
10	100
11	121
12	144
8	64
9	81
$\sum x_i = 88$	$\sum x_i^2 = 804$

$$\frac{\sum x_i}{n} = \frac{88}{10} \quad \text{so} \quad \left(\frac{\sum x_i}{n}\right)^2 = 8.8^2 = 77.44$$

$$\frac{\sum x_i^2}{n} = \frac{804}{10} = 80.4$$

$$\text{Variance} = \frac{\sum x_i^2}{n} - \left(\frac{\sum x_i}{n}\right)^2 = 80.4 - 77.44 = 2.96$$

Standard deviation = $\sqrt{2.96}$ = 1.72.

Summary statistics

Summary statistics are information that gives a brief description of the data.

Tables of summary statistics normally contain some or all of the following:

- Mean
- Mode
- Median
- Maximum value
- Minimum value
- Lower quartile
- Upper quartile
- Standard deviation.

However, there are summary statistics that can be used to calculate other statistics.

Calculating standard deviation from summary statistics

To calculate standard deviation from summary statistics, you would use Σx, Σx^2, n.

2.6 Selecting and critiquing data presentation techniques

Data is often presented in an inappropriate way. Sometimes it is done deliberately to mislead and other times it is done in ignorance of the correct way. One important skill in statistics is to be able to recognise the mistakes and be able correct them. It is also an important skill to be able to choose the best method of presenting a set of data using a graph or chart.

Bar chart

Usually has words along one of the axes rather than numbers.

Bar charts are useful to compare data for two different variables like this

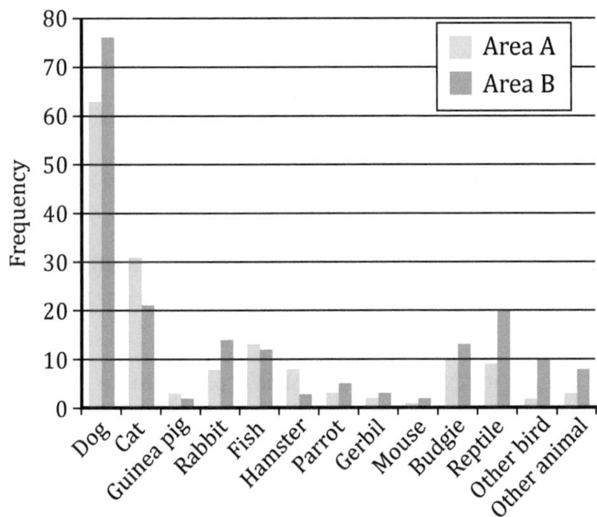

Histogram

Histograms

- Are used to compare the different frequencies of numeric data
- Use data which is continuous so there is no gap between the bars
- Usually have different widths (although they can have the same width).

Data in a form such as this can be used to create a histogram.

The ages (a) of people on the last train	Frequency
$0 \le a < 15$	12
$15 \le a < 20$	24
$20 \le a < 25$	42
$25 \le a < 35$	76
$35 \le a < 50$	12
$50 \le a < 75$	9

Remember to use the data in the table to work out frequency density in order to produce the histogram.

Scatter graph

Scatter graphs:

- Are used to determine if there is any correlation between two quantities
- Are used to see how closely the variables are correlated, if at all
- Are used to look for positive or negative correlation.

Box plot

Box plots:

- Are often used to compare the statistics of two or more sets of data
- Use five values (i.e. minimum, first quartile, median, third quartile, and maximum) to describe a set of data.

Pie chart

Pie charts:

- Can be used to compare distributions (i.e. you can have a pie chart for each distribution)
- Can be used to show percentage or proportional data and usually the percentage represented by each category is provided next to the corresponding slice of pie
- Should only be used for six categories or fewer, because if there are more, it makes it hard for the eye to see the differences between the proportions.

Example

1 Here are three graphs/charts which have been drawn to represent different sets of data. There is a problem with each of the graphs/charts. Describe what the problem is with each one and explain what could be done to improve the graph/chart.

(a)

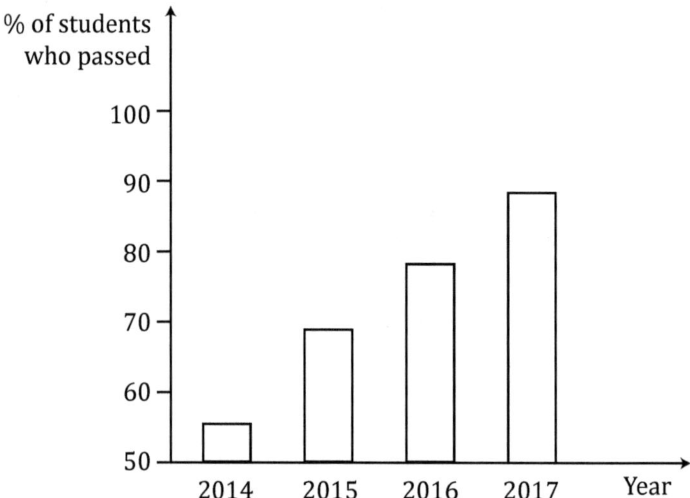

(b)

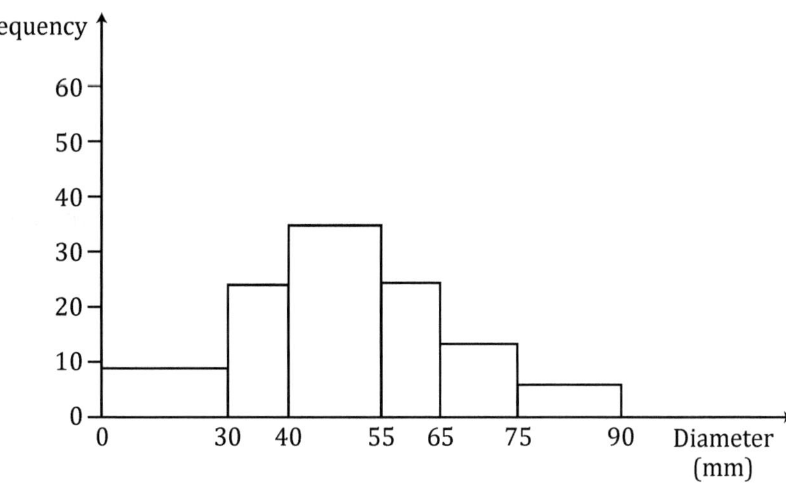

(c)

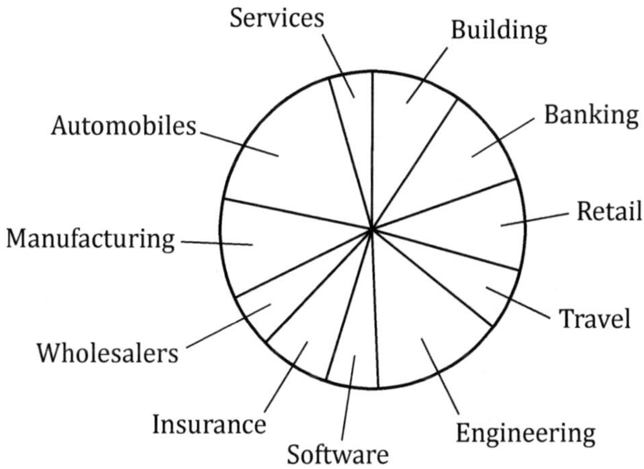

Answer

(a) The vertical axis of the bar chart does not start at zero and it therefore makes it look like the grades have improved considerably from one year to the next. The bar chart should be drawn with the scale on the vertical axis starting from zero.

(b) This is histogram as there are no gaps between the bars and some of the bars are of different widths. Histograms have frequency density plotted on the vertical axis and it is the area and not the height that represents the frequency. The frequency density for each class should be worked out and the graph/chart redrawn.

(c) The pie chart has been drawn with too many sectors and it makes it hard to compare the components. Some of the sectors should be combined and the chart re-drawn so that only 5 to 6 sectors are used.

2.7 Cleaning data (dealing with missing data, errors and outliers)

Dealing with errors

Cleaning data involves detecting and correcting, or removing data as there may be, for example, typos that can be identified and easily corrected. So you would not automatically remove the data point as you would first look to see what the error was. If the data points cannot be corrected, then these data points can be removed and the statistical analysis can be performed on the remaining set of data.

Dealing with outliers

Here is a list of a random sample of household income in a certain road.

£45 000 £50 000 £38 000 £200 000

The mean household income using all of these values is

$$\frac{45\,000 + 50\,000 + 38\,000 + 200\,000}{4} = £83\,250$$

This clearly does not represent the set of data well as the mean is unlike any of the data values. This unrepresentative value is because there is an outlier in the data. An outlier is a value that does not follow the pattern of the data. The £200 000 value has shifted the mean towards a higher value.

If the outlier is removed from the above data the mean household income is

$$\frac{45\,000 + 50\,000 + 38\,000}{3} = £44\,333$$

which is more representative of the remaining data.

Use of the formulae to identify outliers

The £200 000 household income is clearly an outlier but sometimes it is more difficult to spot an outlier. Luckily there is a formula that can be used on data values to decide whether they are outliers or not.

An outlier is any value that is

smaller than $Q_1 - 1.5 \times \text{IQR}$

or larger than $Q_3 + 1.5 \times \text{IQR}$

Examples

1 A random sample of shoe sizes is taken and the following results are obtained.

4, 5, 6, 6, 7, 7, 8, 8, 9, 9, 9, 9, 9, 10, 14

(a) Find the median shoe size.

(b) Find the interquartile range.

(c) Determine whether the data values 4 and 14 are outliers.

Answer

1 (a) The median divides the data in two halves and is the $\frac{n+1}{2}$th value.

As there are 15 values the median is the $\frac{15+1}{2}$ = 8th value

Median shoe size = 8

Always check that the data is in numerical order before finding the median.

(b) Lower quartile = $\frac{n+1}{4}$th value which is the $\frac{15+1}{4}$ = 4th value.

Lower quartile, $Q_1 = 6$

Upper quartile = $\frac{3(n+1)}{4}$th value which is the $\frac{3(15+1)}{4}$ = 12th value.

Upper quartile, $Q_3 = 9$

$\text{IQR} = Q_3 - Q_1 = 9 - 6 = 3$

(c) To see if shoe size 14 is an outlier we need to determine if it is larger than $Q_3 + 1.5 \times \text{IQR}$

$$Q_3 + 1.5 \times \text{IQR} = 9 + 1.5 \times 3 = 13.5$$

As 14 is larger than this value, 14 is an outlier.

To see if shoe size 4 is an outlier we need to determine if it is smaller than $Q_1 - 1.5 \times \text{IQR}$

$$Q_1 - 1.5 \times \text{IQR} = 6 - 1.5 \times 3 = 1.5$$

4 is not smaller than 1.5 so size 4 is not an outlier.

The points removed can be outliers which are those points that stand out as not following the pattern that can be seen in the data. If a scatter graph is drawn and a couple of points do not follow the trend, then these are outliers which can be ignored.

---

2 A company manufactures robot vacuum cleaners that contain rechargeable batteries. The company test a sample of 50 cleaners to investigate how long in hours the battery lasted before needing to be charged again.

Using this data, the following table of summary statistics have been produced.

Summary statistics

How long in hours the battery lasted before needing charging

Number of hours	N	Mean	Standard deviation	Minimum	Lower quartile	Median	Upper quartile	Maximum
	50	2.4	0.56	1.7	2.4	2.7	3.0	3.5

The company would like to include the maximum battery life as being 3.5 hours. It needs to make sure that this value is not an outlier.

Use the summary statistics in the table to show by calculation that 3.5 hours is not an outlier.

Answer

2 $\text{IQR} = Q_3 - Q_1$ (i.e. upper quartile – lower quartile)

$= 3.0 - 2.4$

$= 0.6$

An outlier is a value larger than $Q_3 + 1.5 \times \text{IQR} = 3.0 + 1.5 \times 0.6 = 3.9$

The maximum value is 3.5 which is less than this so the maximum value of 3.5 is not an outlier.

Missing data

Missing data makes a sample less representative and this can affect the conclusions you draw about the population. Often people answering a questionnaire will leave an answer to a particular question blank because it is too personal.

There are the following ways of dealing with missing data:

Delete the samples with any missing data elements

You simply delete the sample with missing data elements. In a small sample this can reduce the sample size so the remaining sample may be no longer representative of the population.

Impute the value of the missing data (put a value in for the missing data items)

There are three main ways this can be done:

- You can substitute the missing data with values from another similar sample.
- You can use the mean from all the other values of the same statistic. For example if the average response to how many days do you drink alcohol per week is two then two can be substituted in for the missing response.
- You can use regression techniques to predict the value of the missing data element based on the relationship between that variable and other variables.

Remove a variable

Suppose a particular question in a questionnaire has a high incidence of missing data, then you could consider removing the variable (i.e. remove the question).

Examples

Most of the questions on this topic require knowledge from most parts of the topic and here are a few examples to show you the sorts of questions you will be asked.

1 A researcher wishes to investigate the relationship between the amount of carbohydrate and the number of calories in different fruits. He compiles a list of 90 different fruits, e.g. apricots, kiwi fruits, raspberries.

As he does not have enough time to collect data for each of the 90 different fruits, he decides to select a simple random sample of 14 different fruits from the list. For each fruit selected, he then uses a dieting website to find the number of calories (kcal) and the amount of carbohydrate (g) in a typical adult portion (e.g. a whole apple, a bunch of 10 grapes, half a cup of strawberries). He enters these data into a spreadsheet for analysis.

(a) Explain how the random number function on a calculator could be used to select this sample of 14 different fruits. [3]

(b) The scatter graph represents 'Number of calories' against 'Carbohydrate' for the sample of 14 different fruits.

(i) Describe the correlation between 'Number of calories' and 'Carbohydrate'. [1]

(ii) Interpret the correlation between 'Number of calories' and 'Carbohydrate' in this context. [1]

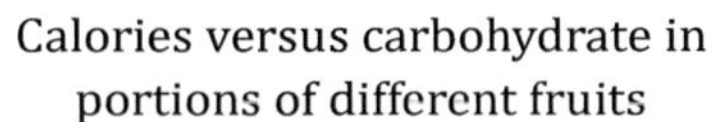
Calories versus carbohydrate in portions of different fruits

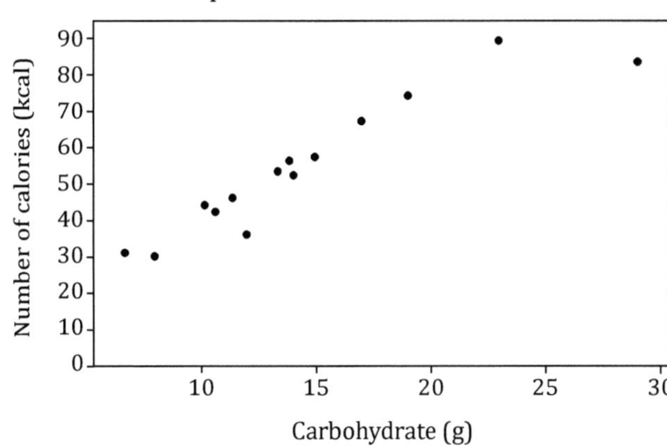

(c) The equation of the regression line for this data set is:

'Number of calories' = 12.4 + 2.9 × 'Carbohydrate'

(i) Interpret the gradient of the regression line in this context. [1]

(ii) Explain why it is reasonable for the regression line to have a non-zero intercept in this context. [1]

Answer

1 (a) Number each type of fruit from 1 to 90. Use a random number generator (on a calculator, website, spreadsheet software, etc.) and use it to pick random numbers from 1 to 90 to choose the fruit. If a number/fruit has already been selected, then select again. Repeat this until 14 different fruit have been obtained for the sample.

(b) (i) Strong positive correlation

(ii) The more carbohydrate a fruit contains the more calories in it.

(c) (i) The number of kilocalories per gram of carbohydrate. As the gradient is 2.9 it means on average the number of kilocalories increases by 2.9 per gram increase in carbohydrate.

(ii) This means that even if there was no carbohydrate in the fruit it would still have some calories from other constituents (e.g. fat).

---

2 Gareth has a keen interest in pop music. He recently read the following claim in a music magazine.

In the pop industry most songs on the radio are not longer than three minutes.

(a) He decided to investigate this claim by recording the lengths of the top 50 singles in the UK Official Singles Chart for the week beginning 17 June 2016. (A 'single' in this context is one digital audio track.)

Comment on the suitability of this sample to investigate the magazine's claim. [1]

Note the use of brackets here for the class intervals. 2.5–(3.0) means you can have a value of from 2.5 to under 3. Note that this does not include 3, as 3 is in the next class.

(b) Gareth recorded the data in the table below.

Length of singles for top 50 UK Official Chart singles 17 June 2016

2.5–(3.0)	3.0–(3.5)	3.5–(4.0)	4.0–(4.5)	4.5–(5.0)	5.0–(5.5)	5.5–(6.0)	6.0–(6.5)	6.5–(7.0)	7.0–(7.5)
3	17	22	7	0	0	0	0	0	1

He used these data to produce a graph of the distributions of the lengths of singles

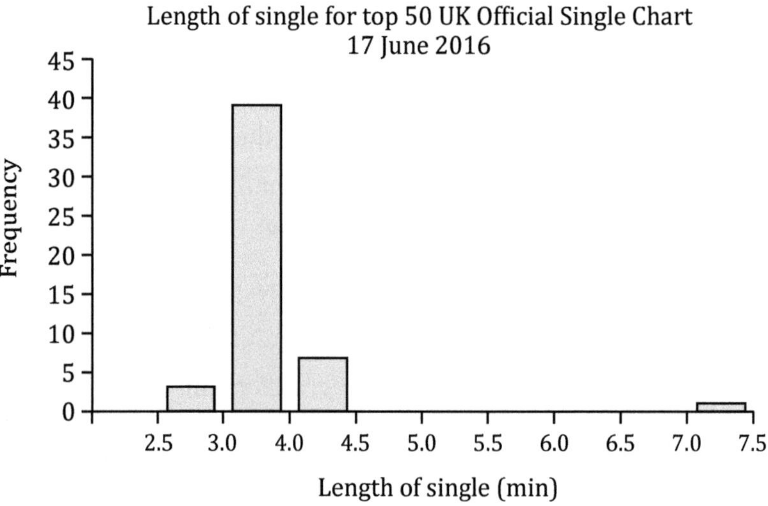

State two corrections that Gareth needs to make to the histogram so that it accurately represents the data in the table. [2]

(c) Gareth also produced a box plot of the lengths of singles.

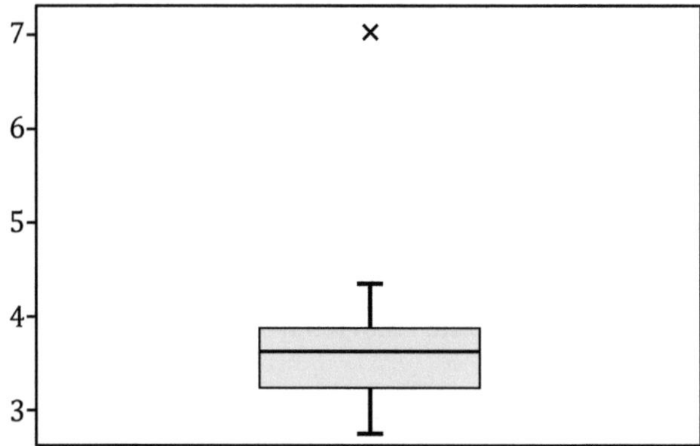

He sees that there is one obvious outlier.

(i) What will happen to the mean if the outlier is removed?

(ii) What will happen to the standard deviation if the outlier is removed? [2]

(d) Gareth decided to remove the outlier. He then produced a table of summary statistics.

(i) Use the appropriate statistics from the table to show, by calculation, that the maximum value for the length of a single is not an outlier.

Summary statistics

Length of single for top 50 UK Official Singles Chart (minutes)

	N	Mean	Standard deviation	Minimum	Lower quartile	Median	Upper quartile	Maximum
Length of single	49	3.57	0.393	2.77	3.26	3.60	3.89	4.38

(ii) State, with a reason, if these statistics support the magazine's claim. [4]

(e) Gareth also calculated summary statistics for the lengths of 30 singles selected at random from his personal collection.

Summary statistics

Length of single for Gareth's random sample of 30 singles (minutes)

	N	Mean	Standard deviation	Minimum	Lower quartile	Median	Upper quartile	Maximum
Length of single	30	3.13	0.364	2.58	2.73	2.92	3.22	3.95

Compare and contrast the distribution of lengths of singles in Gareth's personal collection with the distribution in the top 50 UK Official Singles Chart. [3]

Answer

2 (a) You cannot be sure that the Singles Chart is representative of the population without knowing how the chart is constructed.

(b) The length of a single is a continuous variable so the gaps between the bars should be closed.

The scale on the x-axis should have equal divisions. There is a discrepancy in the scale for 3–4 as it should go 3–3.5 and then 3.5–4.

(c) (i) The value of the outlier is higher than other values so its removal will cause the mean to decrease.

(ii) The spread of the rest of the data will decrease so the standard deviation will decrease.

(d) (i) To see if the maximum length 4.38 min is an outlier we need to determine if it is larger than $Q_3 + 1.5 \times \text{IQR}$

$Q_3 + 1.5 \times \text{IQR} = 3.89 + 1.5 \times (3.89 - 3.26) = 4.84$
(note this is rounded to 3 s.f.)

As the maximum length (4.38 min) is not larger than this value, it is not an outlier.

(ii) Median = 3.60 (from table) so this means that at least half the singles are longer than 3 minutes. This does not support the magazine's claim.

(e) Gareth's singles are shorter than the chart singles on average as shown by the lower mean and median.

Gareth's singles have less spread as shown by the lower range, IQR and standard deviation.

Chart singles have a roughly symmetrical distribution whereas in Gareth's sample of singles, more than half of them are shorter than the mean length.

Test yourself

1 Statistics were gathered on students' marks in GCSE chemistry and physics mock exams. The pairs of marks for each student are summarised in the table below.

Chemistry mark (%)	Physics mark (%)
23	20
35	37
82	90
34	28
40	41
80	76
12	10
13	12
27	30
6	10
21	30
19	21
33	30
51	47
83	90
78	65
80	74
90	85
40	67
56	50

The data in the table was used with computer software to produce the following scatter diagram.

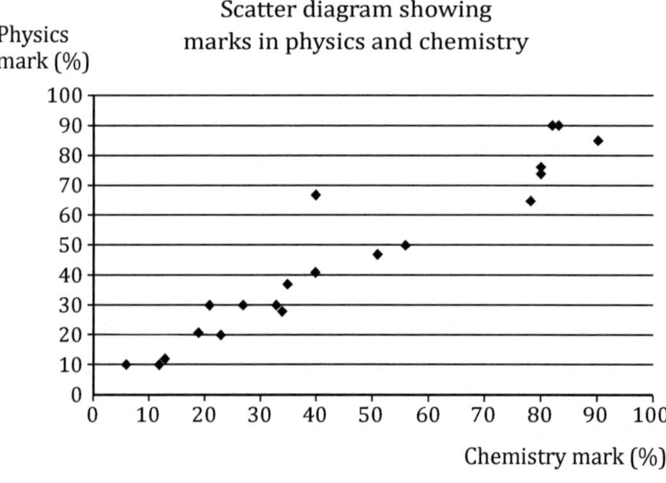

(a) Using the scatter diagram shown above, comment on the correlation between the marks for physics and chemistry.

(b) Explain using the scatter diagram as an example the term 'interpolation'.

(c) The regression line for the data is drawn and its equation is found. Explain the suitability of using the equation to determine the likely mark in physics for a student who obtained a mark of 100% in chemistry.

2 The table below gives information about the weight in kilograms of 151 adult British Bulldogs.

Weight of adult British Bulldog (w kg)	Frequency
$19 < w \leq 20$	2
$20 < w \leq 21$	10
$21 < w \leq 22$	15
$22 < w \leq 23$	20
$23 < w \leq 24$	25
$24 < w \leq 25$	30
$25 < w \leq 26$	26
$26 < w \leq 27$	18
$27 < w \leq 28$	5

This data was used to produce the following cumulative frequency curve.

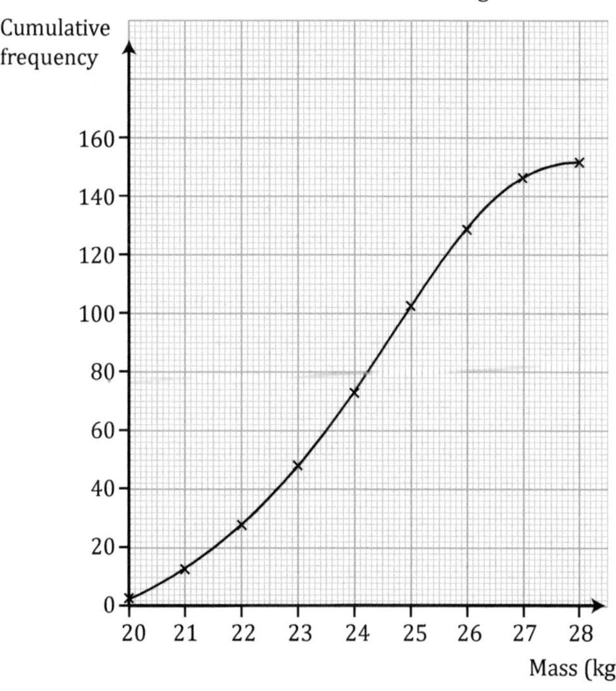

(a) Use the cumulative frequency diagram to work out the following:
   (i) The lower quartile
   (ii) The upper quartile
   (iii) The median

(b) Find the interquartile range and explain the significance of it.

3 A student carried out an experiment to see how the length of a spring varied with the masses placed on the end of it.
A mass of 50 g was hung from the end of the spring and the corresponding length in cm noted. A further 50 g was added and the new length noted. This was repeated until 20 pairs of values were obtained.
The pairs of values were plotted on a graph with the mass in g on the x-axis and the length of the spring in cm plotted on the y-axis. A line of best fit was drawn and the regression equation found.
The regression equation was found to be

Length (cm) = 20 + 0.025 × mass in g

(a) Explain the significance of the intercept on the vertical axis.
(b) A 100 g mass is attached to the spring. Find the length of the spring and its extension.
(c) A mass of 5 kg is to be attached to the spring. Explain why the regression equation should not be used to find the length of the spring.

4 Here are some statements about correlation and causation. You have to decide whether each statement is true or false and then give a reason for your decision.
(a) A scatter diagram is drawn to show latitudes and their corresponding temperatures at various positions north of the equator. If there are 30 different latitudes used, there will be 30 data plots on the diagram.
(b) Causation also implies correlation.
(c) If two variables are positively correlated, then a high value of one of the correlated variables will correspond to a low value of the other.
(d) An investigation revealed that when there were organised firework displays in an area, the number of firework-related accidents decreased. This means that in an area where there were lots of firework-related accidents, there were few organised displays.
(e) If there is positive correlation between arm length and height, then if you are the tallest person in your class you will have the longest arm length.
(f) If two variables are negatively correlated, high values for one variable would suggest a lower value for the other.
(g) In order to test a new drug, half the patients were given a pill that contained no drug and the other half were given the pill containing the new drug. If the conditions of both sets of patients became different then there is causation between the drug and the condition.

5 The number of eggs (x) in 20 nests of woodpeckers is recorded.
The summary data values are as follows:

$$\sum x_i = 102$$

$$\sum x_i^2 = 580$$

Calculate the variance and the standard deviation for this set of data giving both answers correct to 3 significant figures.

6 The masses in kg of 20 male birds of a certain species were recorded.
$\sum x_i = 92$ and $\sum x_i^2 = 435.42$.
(a) Calculate the mean and standard deviation for the masses of the birds.
(b) One of the masses of the birds is 6.8 kg. It is thought that this value could be an outlier.

Summary statistics

Masses of a certain species of bird (kg)

Mass of bird	N	Mean	Standard Deviation	Mini-mum	Lower quartile	Median	Upper quartile	Maxi-mum
	20			3.8	4.05	4.40	5.08	6.8

Using statistics from the above table and your answer to part (a), determine whether this value is an outlier.

(c) The smallest value for the mass of a bird is suspected of being an outlier. Using statistics in the table what is the smallest mass a bird can be without it being an outlier?

7 Data was collected about the number of calories in a cup of breakfast cereal and the number of grams of sugar. A scatter diagram was drawn using the results.

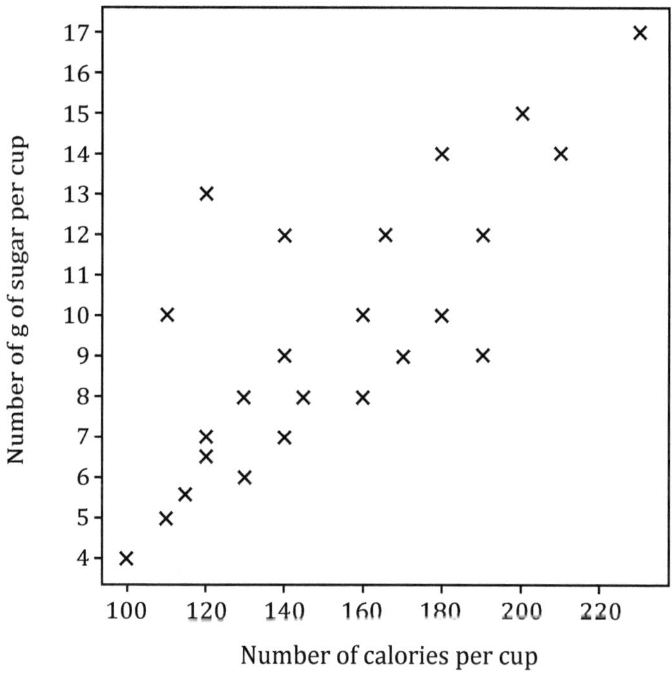

(a) (i) Comment on the correlation between the 'calories per cup' and 'grams of sugar per cup'.

(ii) Interpret the correlation between 'calories per cup' and 'grams of sugar per cup'.

(b) The regression equation for this dataset is
'Grams of sugar per cup' = 0.114 × 'Number of calories per cup'+ 3.50

(i) Interpret the gradient of the regression equation for this model.

(ii) State, giving a reason, whether the regression model could be used to predict the number of grams of sugar per cup for a cup of a new chocolate cereal with 350 calories per cup.

(iii) State giving a reason, whether the relationship between the 'calories per cup' and 'grams of sugar per cup' is causal.

Summary

Check you know the following facts:

Histograms

There are no gaps between the bars and the height of each bar is the frequency density.

The area of the bar is equal to the frequency, where

Frequency (area of bar) = frequency density × class width

Box plots

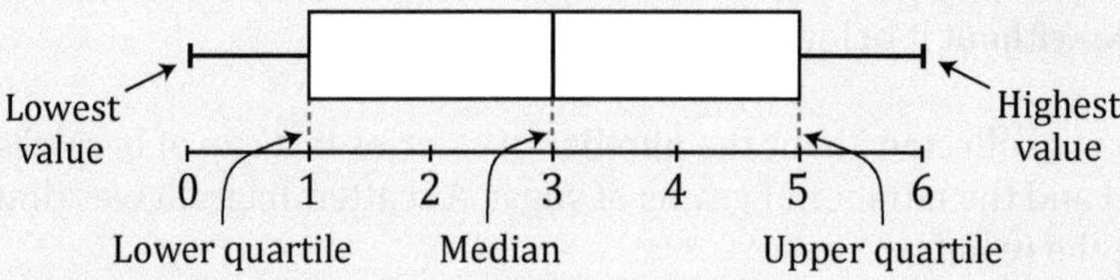

Calculation of the mean

For a **set of values**, mean, $\mu = \frac{\sum x_i}{n}$, where $\sum x_i$ is the sum of all the individual values and n is the number of values

For a **frequency distribution**, mean, $\mu = \frac{\sum f x_i}{\sum f}$, where $\sum f x_i$ is the sum of all the x values, each multiplied by its frequency, f, and $\sum f$ is the sum of all the frequencies.

Working out the mode – the mode is the value(s) or class that occurs most often.

Working out the median – the median is the middle value when the data values are put in order of size. The median is at the $\frac{n+1}{2}$ value.

Measures of central variation (variance, standard deviation, range and interquartile range)

The following measures of the spread of data can be used:

Range

The difference between the largest value and the smallest value in a set of data.

Interquartile range (IQR)

The difference between the upper quartile (Q_3) and the lower quartile (Q_1)

So $\quad \text{IQR} = Q_3 - Q_1$

Variance

Variance $= \frac{\sum (x_i - \mu)^2}{n}$ where $\sum (x_i - \mu)^2$ is the sum of the squares of the differences between each value in the set and the mean μ and n is the total number of values.

The **simplified formula for variance** is, Variance $= \frac{\sum x_i^2}{n} - \left(\frac{\sum x_i}{n}\right)^2$ where $\frac{\sum x_i^2}{n}$ is the mean of the squares of the values and $\left(\frac{\sum x_i}{n}\right)^2$ is the square of the mean of the values.

Standard deviation (σ)

The standard deviation (σ) is the square root of the variance so

$$\sigma = \sqrt{\frac{\sum (x_i - \mu)^2}{n}} \quad \text{or} \quad \sigma = \sqrt{\frac{\sum x_i^2}{n} - \left(\frac{\sum x_i}{n}\right)^2}$$

3 Probability

Introduction

Probability is an important part of statistics as it is concerned with the likelihood of a certain event or events happening. It is used to assess risk when working out the premium for an insurance policy or assessing the risks associated with a business decision. It also is relevant to the many forms of games involving speculation.

You will have come across some probability at GCSE level and this topic reinforces and builds on this prior knowledge. If you feel a bit rusty on probability it will be best to look back over your GCSE notes or a GCSE revision guide before starting this topic.

This topic covers the following:

3.1 Random experiments

3.2 The use of sample space

3.3 Events

3.4 Venn diagrams

3.5 Probability and outcomes

3.6 The addition law for mutually exclusive events

3.7 The generalised addition law

3.8 Multiplication law for independent events

3.1 Random experiments

A random experiment is an experiment, trial, or observation that can be repeated numerous times under the same conditions. The **outcome** of an individual random experiment must in no way be affected by any previous outcome and cannot be predicted with certainty.

Examples of random experiments include:

- The tossing of a coin. The experiment can yield two possible outcomes, heads or tails.
- The roll of a die. The experiment can yield six possible outcomes and these outcomes are the numbers 1 to 6 as the die faces are labelled.
- The selection of a numbered ball (1–50) in a bag. The experiment can yield 50 possible outcomes.

3.2 The use of sample space

A complete list of all possible outcomes of a random experiment is called the **sample space** and is denoted by S.

If a single dice is thrown then there are 6 possible scores, so the sample space is

$$S = \{1, 2, 3, 4, 5, 6\}$$

If two dice are thrown then there are 36 possibilities and the sample space can be written as follows:

$$S = \begin{Bmatrix} 1,1 & 1,2 & 1,3 & 1,4 & 1,5 & 1,6 \\ 2,1 & 2,2 & 2,3 & 2,4 & 2,5 & 2,6 \\ 3,1 & 3,2 & 3,3 & 3,4 & 3,5 & 3,6 \\ 4,1 & 4,2 & 4,3 & 4,4 & 4,5 & 4,6 \\ 5,1 & 5,2 & 5,3 & 5,4 & 5,5 & 5,6 \\ 6,1 & 6,2 & 6,3 & 6,4 & 6,5 & 6,6 \end{Bmatrix}$$

You can draw the sample space without the curly brackets but you should put each pair of values in brackets so the first line could be written like this:

(1,1) (1,2) (1,3) (1,4) (1,5) (1,6)

3.3 Events

An event is a property associated with the outcomes of a random experiment. It is represented by a subset of the sample space.

Examples

1 A cubical die is thrown once. Possible events are:

(a) A, the score obtained is an odd number

(b) B, the score obtained is greater than 4.

For each event, list the elements in the subset of the sample space.

Answer

1 Sample space = set of possible outcomes

= {1, 2, 3, 4, 5, 6}

(a) $A = \{1, 3, 5\}$

(b) $B = \{5, 6\}$

2 A card is selected from a pack of playing cards and the suit is noted. Possible events are:

(a) A, a heart will be selected

(b) B, a red card will be drawn

(c) C, a black card will be drawn.

For each event, list the elements in the subset of the sample space.

Answer

2 Sample space = {heart, diamond, spade, club}

(a) A = {heart}

(b) B = {heart, diamond}

(c) C = {spade, club}

The complement of an event

The complement of an event A is denoted by A', being the event that A does not occur.

In Example 1,

S = {1,2, 3, 4, 5, 6}

A = {1, 3, 5}

and A' = {2, 4, 6}

Elements in S but not in A.

In Example 2,

S = {heart, diamond, spade, club}

B = {heart, diamond}

and B' = {spade, club}

Elements in S but not in B.

Combined events

For two events A and B, the event that A or B or both occur is called the **union** of A and B and is written $A \cup B$.

The event that both A and B occur is called the **intersection** of A and B and is written as $A \cap B$.

In Example 1, A = {1, 3, 5}, B = {5, 6}

and $A \cup B$ = {1, 3, 5, 6}

and $A \cap B$ = {5}

In Example 2, A = {heart}, B = {heart, diamond}, C = {spade, club},

and $A \cup C$ = {heart, spade, club}

and $A \cap B$ ={heart}

Subsets

Suppose $A = \{1, 3, 5, 7, 9, 11, 13\}$ and $B = \{1, 5, 11, 13\}$ then all the members of set B are also members of set A. We say that B is a **subset** of A. Using set symbols, this is written as

$$B \subset A$$

3.4 Venn diagrams

The diagram below shows a Venn diagram. The rectangle shown below, represents the sample space S and inside this are two events A and B represented by circles.

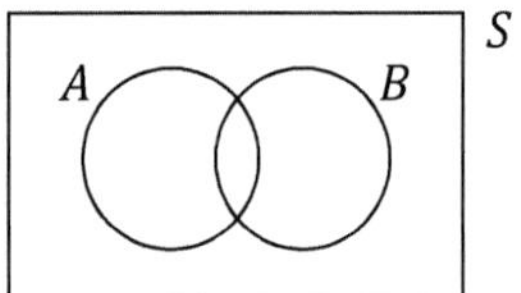

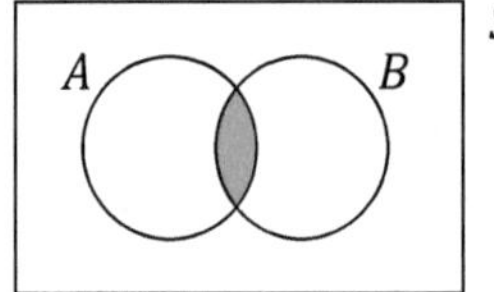

This shaded region represents the event where both A and B occur and is $A \cap B$.

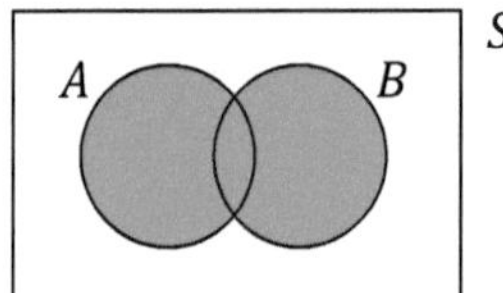

This shaded region represents the event that A or B, or both A and B occur and is $A \cup B$.

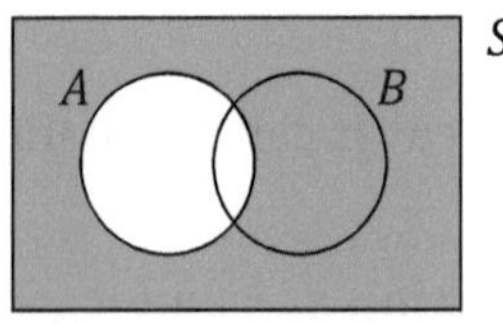

On the Venn diagram, A' is shown by the shading of the region which is *not* in A.

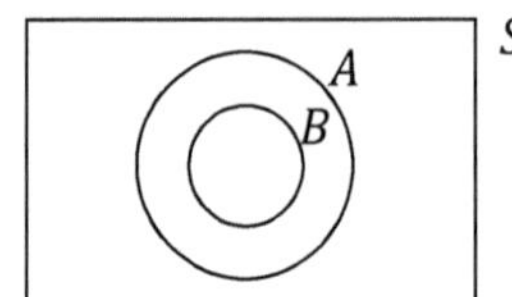

B is a subset of A, that is $B \subset A$.

$B \subset A$ means that all the contents of set B are also in the larger set A.

It is important to be able to describe regions using symbols and vice versa.

Examples

1 Describe the shaded region using symbols for each of the following Venn diagrams:

(a)

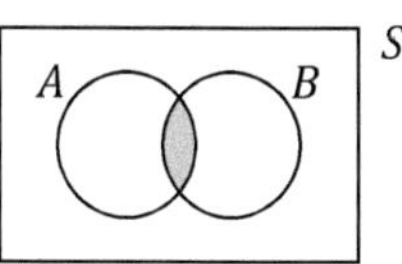

(b)

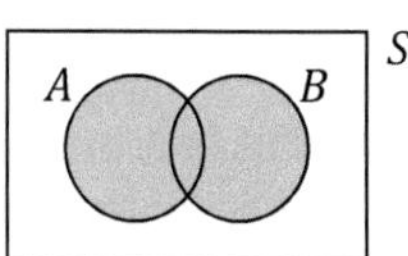

(c)

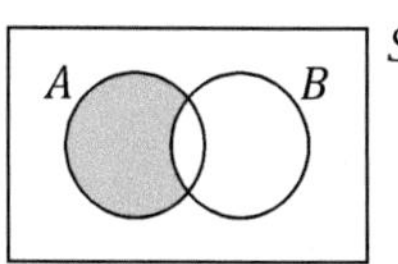

(d)

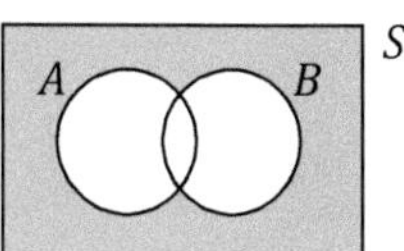

Answer

1 (a) $A \cap B$

(b) $A \cup B$

(c) $A \cap B'$

(d) $(A \cup B)'$

---

2 Draw Venn diagrams with events A and B and shade in the regions represented by each of the following:

(a) B'

(b) $A' \cap B$

(c) $A' \cap B'$

Answer

2 (a)

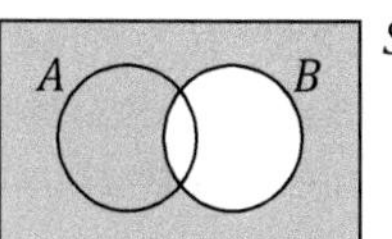

(b)

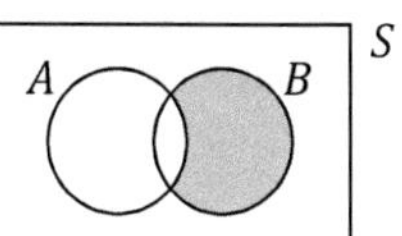

(c)

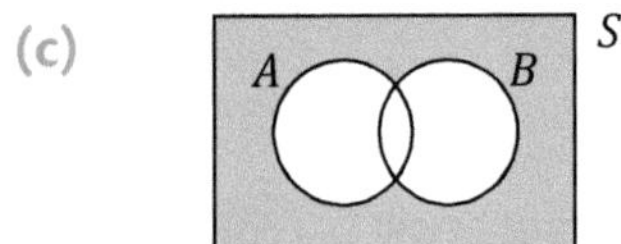

3.5 Probability and outcomes

Associated with an outcome (or event) of a random experiment is a measure of the certainty with which it can occur. This measure is known as the probability of the outcome (event). There are two ways in which probability is defined.

Equally likely outcomes

For any outcome, $r = 1$
Prob (outcome) $= \frac{1}{N}$

In some random experiments, it appears that any outcome is no more likely to occur than any other. In such a case, if there are N equally likely outcomes and r of these outcomes favour the event A, then

$$\text{Probability } (A \text{ occurs}) = \frac{r}{N}$$

Example

4 aces in a pack of 52 possible outcomes.

When a card is drawn from a well-shuffled pack of 52 playing cards, the probability an ace is drawn $= \frac{4}{52}$

Non-equally likely outcomes (relative frequency method)

Sometimes it cannot be assumed that outcomes are equally likely. In such a case it is supposed that a large number of trials N of the random experiment are performed and the event A occurs $r(A)$ times.

Then we take

$$\text{estimate of probability } (A \text{ occurs}) = \frac{r(A)}{N}$$

Thus, for example, if a damaged cubic die is thrown 1960 times and a score of greater than 4 is obtained 256 times we estimate,

$$\text{probability (score} > 4) = \frac{256}{1960} = 0.13 \text{ (correct to 2 d.p.)}$$

Whichever method is used to define probability, we are able to define some rules to enable us to solve problems. Before doing so, we note if we regard the complete sample space as an event, then $P(S) = 1$. Similarly, for an event that cannot occur, i.e. is not contained in the sample space, the probability is zero.

Similarly for circle B.

We use Venn diagrams to establish the laws of probability, it being understood that the area enclosed by the circle for A is the probability of A occurring.

$P(A')$

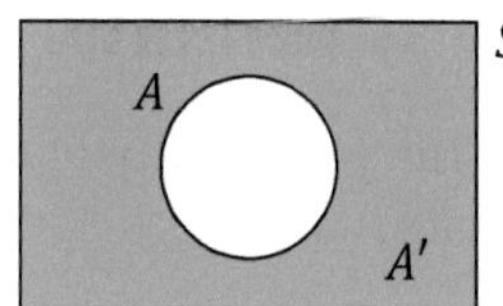

Since $P(S) = 1$

$$P(A') = 1 - P(A)$$

3.6 The addition law for mutually exclusive events

If events A and B are mutually exclusive, it means that event A can happen or event B can happen but they cannot both happen. On the Venn diagram, you can see that there is no intersection between A and B.

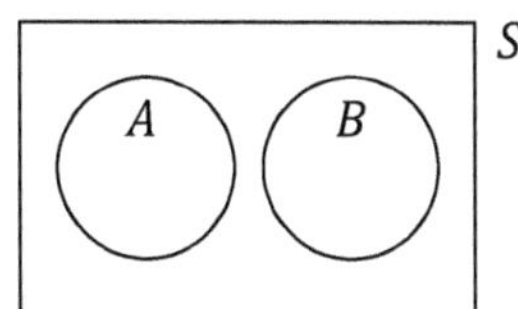

If you want the probability of A or B happening, you simply add the probability of A occurring to the probability of B occurring. This can be written as follows:

$$P(A \cup B) = P(A) + P(B)$$

This formula only applies to mutually exclusive events and is not included in the formula booklet and needs to be remembered. Note that this can be used to prove that two events A and B are mutually exclusive.

Example

1 Two events A and B are such that

$$P(A) = 0.35,\ P(B) = 0.45 \text{ and } P(A' \cap B') = 0.6$$

Determine whether events A and B are mutually exclusive.

Answer

1

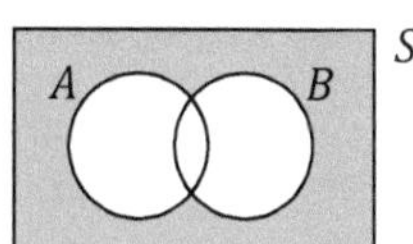

$$P(A \cup B) = 1 - P(A' \cap B') = 1 - 0.6 = 0.4$$

Now if the events were mutually exclusive we can write

$$P(A \cup B) = P(A) + P(B) = 0.35 + 0.45 = 0.8$$

As these two results do not agree, the two events are not mutually exclusive.

One way to solve this is by drawing a Venn diagram for non mutually exclusive events. Note that if the events were mutually exclusive there would be no overlap between events A and B. The shaded region shows $A' \cap B'$ (i.e. everything that is neither A nor B). You can see that the region not shaded represents $A \cup B$. Remember that the total probability of the sample space S is 1.

3.7 The generalised addition law

Adding areas of A and B involves adding $P(A \cap B)$ twice, one of which must be subtracted.

The generalised addition law links the probability of the intersection with the probability of the union of two events A and B.

Then $P(A \cup B) = P(A) + P(B) - P(A \cap B)$.

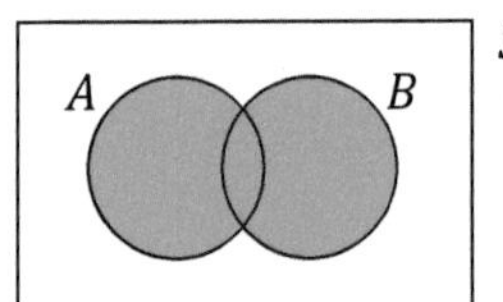

This is a very important formula which must be remembered.

3.8 Multiplication law for independent events

This formula is not included in the formula booklet and will need to be remembered. Note also that this formula only applies to independent events. Note that $P(A \cap B)$ represents the probability of events A and B both occurring. Students are often confused between independent events and mutually exclusive events. Note the difference.

When an event has no effect on another event, they are said to be independent events. For example, if event A occurs then it will have no effect on event B happening and vice versa.

The multiplication law for independent events is as follows:

$$P(A \cap B) = P(A) \times P(B)$$

Examples

1 Events A and B are such that

$P(A) = 0{\cdot}3, \quad P(B) = 0{\cdot}2, \quad P(A \cup B) = 0{\cdot}44.$

(a) Show that A and B are independent.

(b) Calculate the probability of exactly one of the two events occurring.

Answer

This is the generalised addition law and it can be looked up in the formula booklet.

1 (a) $P(A \cup B) = P(A) + P(B) - P(A \cap B)$

Rearranging, we obtain

$$\begin{aligned} P(A \cap B) &= P(A) + P(B) - P(A \cup B) \\ &= 0.3 + 0.2 - 0.44 \\ &= 0.06 \end{aligned}$$

Note that this multiplication law only applies if the events are independent.

If events are independent $P(A \cap B) = P(A) \times P(B)$

$$\begin{aligned} &= 0.3 \times 0.2 \\ &= 0.06 \end{aligned}$$

Hence the events are independent because $P(A \cap B) = P(A) \times P(B)$

(b) Probability of exactly one event $= P(A \cup B) - P(A \cap B) = 0.44 - 0.06 = 0.38$

Another way to work this out is as follows:

$P(A \text{ only}) = 0.3 - 0.06 = 0.24$

$P(B \text{ only}) = 0.2 - 0.06 = 0.14$

$P(A \text{ or } B \text{ only}) = 0.24 + 0.14$
$= 0.38$

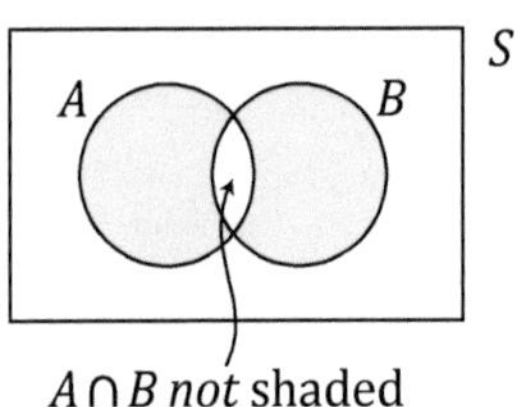

$A \cap B$ *not* shaded

Alternative answer

This problem could also be solved as follows:

1 (a)

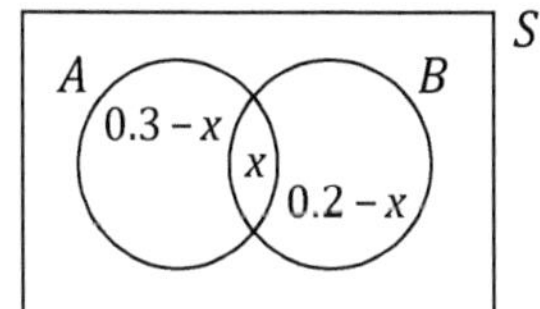

Let $x = P(A \cap B)$ so the probability of A only is $0.3 - x$ and the probability of B only is $0.2 - x$.

From the question, $P(A \cup B) = 0.44$, so from the Venn diagram we have

$$0.3 - x + x + 0.2 - x = 0.44$$

$$0.5 - x = 0.44$$

Hence $x = 0.06$

If the two events A and B are independent, then

$$P(A \cap B) = P(A) \times P(B) = 0.3 \times 0.2 = 0.06$$

As $P(A \cap B) = 0.06$ and $P(A) \times P(B) = 0.06$, we have $P(A \cap B) = P(A) \times P(B)$ which proves that events A and B are independent.

(b) Probability of exactly one of the two events occurring is the shaded area shown.

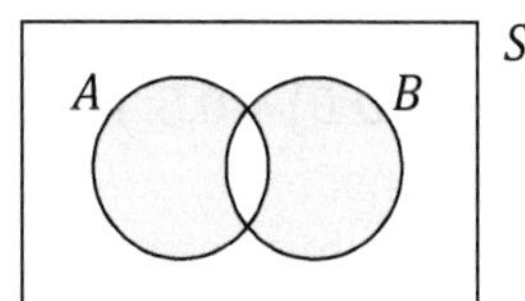

P(exactly one of the events occurring) $= 0.3 - x + 0.2 - x = 0.5 - 2x$

As $x = 0.06$, required probability $= 0.5 - 0.12 = 0.38$

BOOST Grade

There are often several ways of solving the same problem. Many people find a visual method of solving a problem like this easier. Probability questions can often be solved using diagrams such as Venn diagrams and tree diagrams. Choose the method you find easiest.

---

2 The independent events A and B are such that

$P(A) = 0{\cdot}6, \quad P(B) = 0{\cdot}3.$

Find the value of

(a) $P(A \cup B)$ [3]

(b) $P(A \cup B')$ [3]

Answer

2 (a) Using $P(A \cup B) = P(A) + P(B) - P(A \cap B)$, we obtain

$$P(A \cup B) = 0.6 + 0.3 - (0.6 \times 0.3) = 0.72$$

(b) $P(B') = 1 - P(B) = 1 - 0.3 = 0.7$

Now, $P(A \cup B') = P(A) + P(B') - P(A \cap B')$

$= P(A) + P(B') - P(A) \times P(B')$

$= 0.6 + 0.7 - 0.6 \times 0.7$

$= 0.88$

Note that

$P(A \cap B) = P(A) \times P(B)$
$= 0.6 \times 0.3 = 0.18$

Note you can use the result

$P(A \cup B) = P(A) + P(B) - P(A \cap B)$

but replacing the occurrences of B with B' to obtain the following:

$P(A \cup B') = P(A) + P(B') - P(A \cap B')$

Note A and B are independent.

This part could also be solved by drawing a Venn diagram.

Alternative method

2 (a) As the events are independent $P(A \cap B) = P(A) \times P(B)$

$= 0.6 \times 0.3 = 0.18$

$P(A \text{ only}) = 0.6 - 0.18 = 0.42$

$P(B \text{ only}) = 0.3 - 0.18 = 0.12$

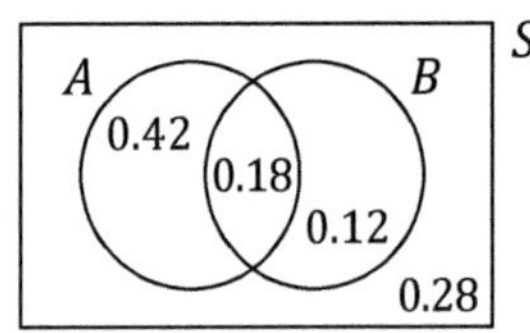

$P(A \cup B) = 0.42 + 0.18 + 0.12 = 0.72.$

The Venn diagram is drawn with the shaded region representing $A \cup B'$.

It is then easy to spot that if you subtract the probability of event B only from 1, you obtain the required probability.

(b)

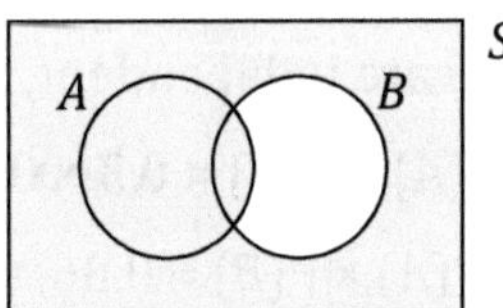

$P(A \cup B') = 1 - P(B \text{ only occurs}) = 1 - 0.12 = 0.88$

3 A and B are two events such that

$P(A) = 0.35$, $P(B) = 0.25$ and $P(A \cup B) = 0.5$. Find

(a) $P(A \cap B)$

(b) $P(A')$

(c) $P(A \cup B')$

Answer

Rearrange
$P(A \cup B) = P(A) + P(B) - P(A \cap B)$

3 (a) $P(A \cap B) = P(A) + P(B) - P(A \cup B)$

$= 0.35 + 0.25 - 0.5$

$= 0.1$

(b) $P(A') = 1 - P(A) = 1 - 0.35 = 0.65$

(c)

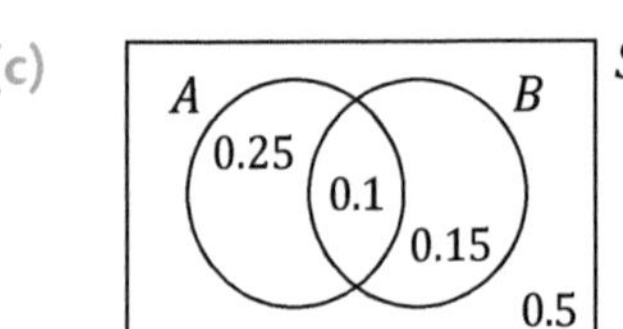

$P(B') = 1 - P(B) = 1 - 0.25 = 0.75$

$P(A \cup B') = P(A) + P(B') - P(A \cap B') = 0.35 + 0.75 - 0.25 = 0.85$

4 The events A and B are such that $P(A) = P(B) = p$ and $P(A \cup B) = 0{\cdot}64$.

(a) Given that A and B are mutually exclusive, find the value of p. [2]

(b) Given, instead, that A and B are independent, show that

$$25p^2 - 50p + k = 0,$$

where k is a constant whose value should be found.

Hence find the value of p. [5]

Answer

4 (a) $p + p = 0.64$

$2p = 0.64$

$p = 0.32$

As A and B are mutually exclusive there is no overlap between events A and B. Hence $P(A \cup B)$ is simply the probability of A or B occurring so the probabilities for each of these events are added.

(b) $P(A \cap B) = p \times p = p^2$

Using $P(A \cup B) = P(A) + P(B) - P(A \cap B)$, we obtain

$0.64 = 2p - p^2$

$p^2 - 2p + 0.64 = 0$

As events A and B are now independent, there may be an overlap between the two events, so $A \cap B$ could exist.

Multiplying the above equation through by 25, we obtain

$25p^2 - 50p + 16 = 0$

Comparing this equation with the one given in the question, we obtain

$k = 16$

Factorising $25p^2 - 50p + 16 = 0$, we obtain

$(5p - 2)(5p - 8) = 0$

So $p = \frac{2}{5}$ or $\frac{8}{5}$

This is a quadratic equation that can be solved by factorising, completing the square or using the formula.

However, as p is a probability, it cannot be greater than 1,

so the answer $p = \frac{8}{5}$ is disregarded.

Hence, $p = \frac{2}{5}$ (or 0.4 as a decimal).

BOOST Grade

When you solve a quadratic you often get two different answers. When this happens always ask yourself if both or just one of the answers is allowable.

5 Two events A and B are such that $P(A) = 0.3$ and $P(B) = 0.6$.

Find the value of $P(A \cup B)$ when

(a) A, B are mutually exclusive

(b) A, B are independent

(c) $A \subset B$

Answer

5 (a) $P(A \cup B) = P(A) + P(B)$

$= 0.3 + 0.6 = 0.9$

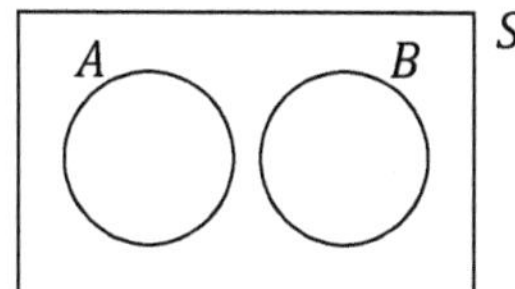

As A and B are mutually exclusive A and B cannot occur so there is no overlap between the sets. Note you are not required to draw any Venn diagrams although you can draw them if it helps you.

> Notice that the overlap would be added twice if you just consider P(A) + P(B). Hence we need to subtract one of these overlaps (i.e. P($A \cap B$)).

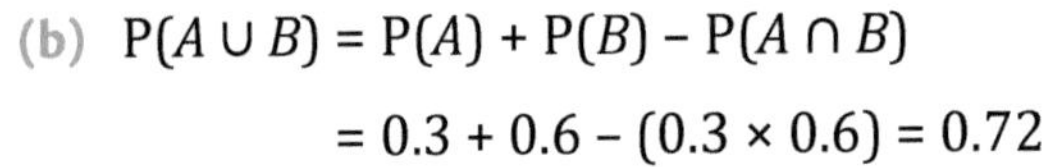

(b) $P(A \cup B) = P(A) + P(B) - P(A \cap B)$

$= 0.3 + 0.6 - (0.3 \times 0.6) = 0.72$

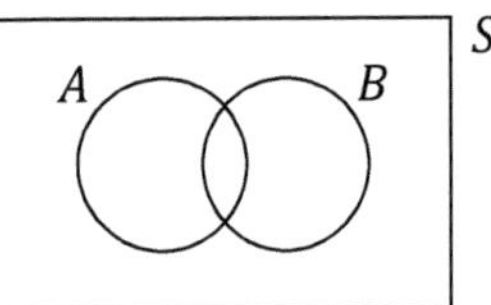

> Set A is included inside set B so P(A) is already included in P(B).

(c) $P(A \cup B) = P(B)$

$= 0.6$

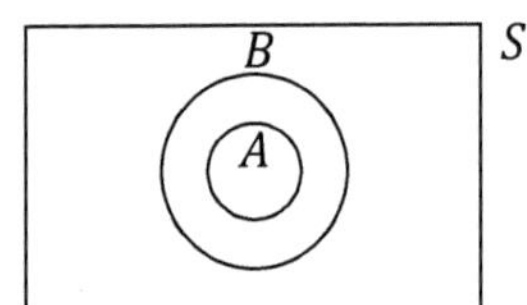

---

6 The independent events A, B are such that P(A) = 0.2, $P(A \cup B) = 0.4$.

(a) Determine the value of P(B). [4]

(b) Calculate the probability that exactly one of the events A, B occurs. [3]

Answer

6 (a) $P(A \cup B) = P(A) + P(B) - P(A \cap B)$

$0.4 = 0.2 + P(B) - P(A \cap B)$ (1)

Now $P(A \cap B) = P(A) \times P(B) = 0.2\,P(B)$

Substituting these values into equation (1) gives

$0.4 = 0.2 + P(B) - 0.2\,P(B)$

$0.2 = 0.8\,P(B)$

$P(B) = 0.25$

(b) You can draw a Venn diagram to show the required region (shown shaded).

This represents the probability of A but not B or the probability of not A but B.

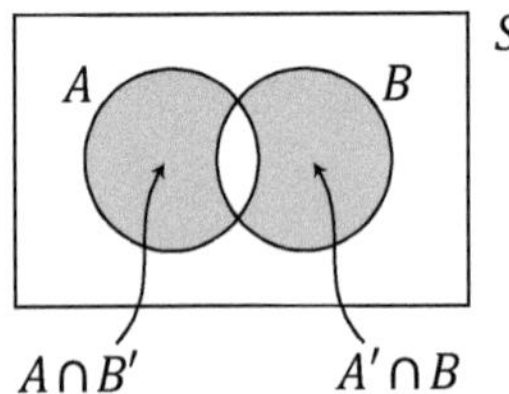

Now $P(B') = 1 - P(B) = 1 - 0.25 = 0.75$

$P(A') = 1 - P(A) = 1 - 0.2 = 0.8$

Required probability $= P(A \cap B') + P(A' \cap B)$

$= 0.2 \times 0.75 + 0.8 \times 0.25$

$= 0.15 + 0.2$

$= 0.35$

Active Learning

Produce a poster containing a summary of using sets to describe probabilities. Your poster should be able to help students understand the use of Venn diagrams to solve probability problems.

Test yourself

1. The events A and B are such that:
   $P(A) = 0.45$, $P(B) = 0.30$, $P(A \cap B) = 0.25$
   Calculate the probability of:
   (a) $P(A \cup B)$
   (b) $P(A' \cap B')$.

2. Amy and Bethany each throw a fair cubical die with faces numbered,
   1, 2, 3, 4, 5, 6.
   (a) Calculate the probability that the score on Amy's die is:
   (i) equal to the score on Bethany's die
   (ii) greater than the score on Bethany's die.
   (b) Given that the sum of the scores on the two dice is 4, find the probability that the two scores are equal.

3. A and B are two independent events such that:
   $P(A) = 0.3$ and $P(A \cup B) = 0.5$.
   (a) Calculate $P(B)$.
   (b) Calculate the probability that exactly one of the two events occurs.

4. The events A and B are such that:
   $P(A) = 0.4$, $P(B) = 0.35$ and $P(A' \cap B') = 0.4$.
   Determine whether:
   (a) A and B are mutually exclusive
   (b) A and B are independent.

5. Two independent events A and B are such that:
   $P(A) = 0.4$, $P(B) = 0.3$
   Find:
   (a) $P(A \cap B)$ [1]
   (b) $P(A \cup B)$ [3]
   (c) the probability that neither A nor B occurs. [3]

6. Two events A and B are such that:
   $P(A) = 0.4$, $P(B) = 0.2$, $P(A \cup B) = 0.5$
   (a) Calculate $P(A \cap B)$. [2]
   (b) Determine whether or not A and B are independent. [2]

7. Events A and B are such that
   $P(A) = 0{\cdot}2$, $P(B) = 0{\cdot}4$, $P(A \cup B) = 0{\cdot}52$.
   (a) Show that A and B are independent. [5]
   (b) Calculate the probability of exactly one of the two events occurring. [2]

8. The events A and B are such that
   $P(A) = 0{\cdot}3$, $P(B) = 0{\cdot}4$.
   Evaluate $P(A \cup B)$ in each of the following cases.
   (a) A and B are mutually exclusive.
   (b) A and B are independent.

9. The events A, B are such that $P(A) = 0.2$, $P(B) = 0.3$. Determine the value of $P(A \cup B)$ when
   (a) A, B are mutually exclusive, [2]
   (b) A, B are independent, [3]
   (c) $A \subset B$. [1]

Summary

Venn diagrams

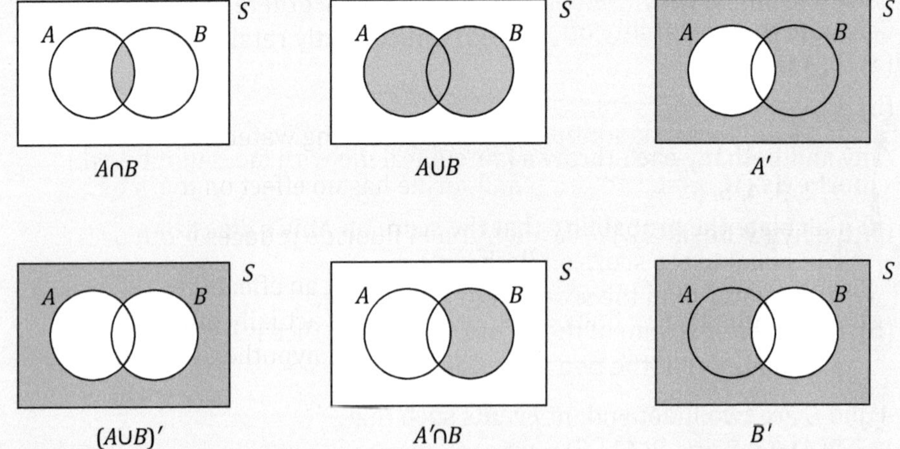

Complementary events

The probability that A does not occur (i.e. A') is one minus the probability that event A does occur.

$$P(A') = 1 - P(A)$$

The meaning of independent events and mutually exclusive events

When an event has no effect on another event, they are said to be independent events. For example, if event A occurs then it will have no effect on event B happening and vice versa.

If events A and B are mutually exclusive then A or B can occur but not both.

The addition law for mutually exclusive events

$$P(A \cup B) = P(A) + P(B)$$

The generalised addition law

$$P(A \cup B) = P(A) + P(B) - P(A \cap B)$$

Multiplication law for independent events

$$P(A \cap B) = P(A) \times P(B)$$

4 Statistical distributions

Introduction

When data is collected it can be modelled as one of the many distributions that can be used to make predictions based on the data. This sounds a bit complicated and it is more complicated to describe than actually do.

This topic covers the following:

4.1 The binomial distribution as a model

4.2 The Poisson distribution as a model

4.3 The discrete uniform distribution as a model

4.4 Selecting an appropriate probability distribution

4.1 The binomial distribution as a model

Bernoulli trials

A Bernoulli trial is an experiment whose outcome is random and can be either of two possible outcomes 'success' or 'failure'.

In order to conduct a Bernoulli trial the question needs to be phrased in such a way that the event can be categorised into success if it takes place or failure if it doesn't.

For example, you could ask the question 'did the die land on a 6?' If it did then this would be considered success and if it didn't, it would be classed as failure. Independent repeated trials of an experiment with two outcomes only are called Bernoulli trials.

The probability of success p added to the probability of failure q will add up to one.

If p is the probability of success in a Bernoulli trial and q is the probability of failure, then we have the result:

$$p + q = 1$$

Binomial distribution

A binomial experiment is closely related to a Bernoulli experiment. It consists of a fixed number of trials, n, each with a probability p of success and it counts the number of successes, x. This is sometimes abbreviated in the following way B(n, p). The binomial distribution is a discrete probability distribution and the probability of a particular number of successes, x, occurring is given by the following formula:

Note that this formula is included in the formula booklet.

$$P(X = x) = \binom{n}{x}p^x(1 - p)^{n-x}$$

Note that $\binom{n}{x}$ can also be written as ${}^nC_x$ so when there are numerical values for n and x, these can be substituted into your calculator for n and x to obtain a value for $\binom{n}{x}$. It is important to note that on many calculators the x is replaced by an r to give ${}^nC_r$.

The probability of failure for each trial is $1 - p = q$.

The binomial distribution formula can only be used with:

- Independent trials (i.e. where the probability of one event does not depend on another).
- Trials where there is a constant probability of success.
- A fixed number of trials.
- Where there is only success or failure.

You are often asked about the conditions for using the binomial distribution in exam questions. You need to remember these conditions.

Use your calculator to work out the following and check you obtain the answers shown.

Examples

1 Work out the following:

(a) $\binom{5}{0}$

(b) $\binom{5}{1}$

(c) $\binom{8}{5}$

(d) $\binom{10}{5}$

Answers

(a) $\binom{5}{0} = 1$

(b) $\binom{5}{1} = 5$

(c) $\binom{8}{5} = 56$

(d) $\binom{10}{5} = 252$

---

2 A salesperson makes 60 calls to potential customers during a particular week. The probability of making a sale at each call is independent of other calls and is 0.3.

Find the probability that during a particular week, he makes:

(a) Exactly 10 sales.

(b) Exactly 19 or 20 sales.

Answer

2 (a) Using $P(X = x) = \binom{n}{x}p^x(1 - p)^{n-x}$

with $x = 10$, $n = 60$ and $p = 0.3$ we obtain

$$P(X = 10) = \binom{60}{10} 0.3^{10}(1 - 0.3)^{60-10}$$

$$= \binom{60}{10} 0.3^{10}(0.7)^{50}$$

$= 0.008$ (correct to 3 decimal places)

Note the assumptions in using the binomial distribution.

(b) Using $P(X = x) = \binom{n}{x}p^x(1 - p)^{n-x}$

with $x = 19$, $n = 60$ and $p = 0.3$ we obtain

$$P(X = 19) = \binom{60}{19} 0.3^{19}(1 - 0.3)^{60-19}$$

$$= \binom{60}{19} 0.3^{19}(0.7)^{41}$$

$= 0.1059$ (correct to 4 decimal places)

Use the formula for the binomial distribution, $P(X = x) = \binom{n}{x}p^x(1 - p)^{n-x}$ which is obtained from the formula booklet.

Using $P(X = x) = \binom{n}{x}p^x(1 - p)^{n-x}$

with $x = 20$, $n = 60$ and $p = 0.3$ we obtain

$$P(X = 20) = \binom{60}{20} 0.3^{20}(1 - 0.3)^{60-20}$$

$$= \binom{60}{20} 0.3^{20}(0.7)^{40}$$

$= 0.0931$ (correct to 4 decimal places)

The probability of obtaining 19 or 20 is found by adding the two probabilities together.

Now, $P(X = 19 \text{ or } 20) = P(X = 19) + P(X = 20)$

$= 0.1059 + 0.0931$

$= 0.199$ (correct to 3 decimal places)

Using binomial distribution tables to determine probabilities

Finding probabilities using the formula is tedious in some situations, but luckily there is a quicker way of finding them than by calculation. Tables of the binomial distribution function can be used. These tables are provided in the exam and you must familiarise yourself in using them.

The binomial distribution function table gives the probability of obtaining at most x successes in a sequence of n independent trials, each of which has a probability of success p.

The table works out the probability for the following function:

This formula determines the probability of less than or equal to x successes.

$$P(X \le x) = \sum_{r=0}^{x} \binom{n}{r} p^r (1-p)^{n-r}$$

To use the tables you firstly need to check carefully that you are using the correct table in the booklet of statistical tables. Make sure you are using Table 1 The Binomial Distribution Function.

There are three quantities that you need in order to obtain the probability in the table, which are:

- n, the total number of trials
- x, the number of successes
- p, the probability of success

For example, suppose that the probability of a particular component being rejected in a batch is 0.3 independently of all other components. What is the probability that in a batch of 10 components, 3 or fewer components are rejected?

Note that in the question, 3 or fewer would include 3. Hence we are finding the probability of 0, 1, 2 or 3 components being rejected.

In this case we have $n = 10$, $p = 0.3$ and $x = 3$

Let us explore the outputs of the tables for these values.

The tables are used in the following way.

- First you have to find the correct table in the booklet. Here you are using the binomial distribution function tables.
- Look for the section which refers to the value of n (i.e. 10 in this case).
- You then look down the column and select the value for x (i.e. 3 in this case).
- Then look at the probability headings at the top of the page to find the required value for p (0.30 in this case). Read off the intersection between the value for x and the value for p and you have the probability. Check that for this example you obtain the probability of 0.6496.

Here is a page of the probability table that you can use to check the answer to the above problem.

BINOMIAL DISTRIBUTION FUNCTION

n	x \ p	0.01	0.02	0.03	0.04	0.05	0.06	0.07	0.08	0.09	0.10	0.15	0.20	0.25	0.30	0.35	0.40	0.45	0.50	p / x
n = 9	0	.9135	.8337	.7602	.6925	.6302	.5730	.5204	.4722	.4279	.3874	.2316	.1342	.0751	.0404	.0207	.0101	.0046	.0020	0
	1	.9966	.9869	.9718	.9522	.9288	.9022	.8729	.8417	.8088	.7748	.5995	.4362	.3003	.1960	.1211	.0705	.0385	.0195	1
	2	.9999	.9994	.9980	.9955	.9916	.9862	.9791	.9702	.9595	.9470	.8591	.7382	.6007	.4628	.3373	.2318	.1495	.0898	2
	3	1.000	1.000	.9999	.9997	.9994	.9987	.9977	.9963	.9943	.9917	.9661	.9144	.8343	.7297	.6089	.4826	.3614	.2539	3
	4			1.000	1.000	1.000	.9999	.9998	.9997	.9995	.9991	.9944	.9804	.9511	.9012	.8283	.7334	.6214	.5000	4
	5						1.000	1.000	1.000	1.000	.9999	.9994	.9969	.9900	.9747	.9464	.9006	.8342	.7461	5
	6										1.000	1.000	.9997	.9987	.9957	.9888	.9750	.9502	.9102	6
	7												1.000	.9999	.9996	.9986	.9962	.9909	.9805	7
	8													1.000	1.000	.9999	.9997	.9992	.9980	8
	9															1.000	1.000	1.000	1.000	9
n = 10	0	.9044	.8171	.7374	.6648	.5987	.5386	.4840	.4344	.3894	.3487	.1969	.1074	.0563	.0282	.0135	.0060	.0025	.0010	0
	1	.9957	.9838	.9655	.9418	.9139	.8824	.8483	.8121	.7746	.7361	.5443	.3758	.2440	.1493	.0860	.0464	.0233	.0107	1
	2	.9999	.9991	.9972	.9938	.9885	.9812	.9717	.9599	.9460	.9298	.8202	.6778	.5256	.3828	.2616	.1673	.0996	.0547	2
	3	1.000	1.000	.9999	.9996	.9990	.9980	.9964	.9942	.9912	.9872	.9500	.8791	.7759	.6496	.5138	.3823	.2660	.1719	3
	4			1.000	1.000	.9999	.9998	.9997	.9994	.9990	.9984	.9901	.9672	.9219	.8497	.7515	.6331	.5044	.3770	4
	5					1.000	1.000	1.000	1.000	.9999	.9999	.9986	.9936	.9803	.9527	.9051	.8338	.7384	.6230	5
	6									1.000	1.000	.9999	.9991	.9965	.9894	.9740	.9452	.8980	.8281	6
	7											1.000	.9999	.9996	.9984	.9952	.9877	.9726	.9453	7
	8												1.000	1.000	.9999	.9995	.9983	.9955	.9893	8
	9														1.000	1.000	.9999	.9997	.9990	9
	10																1.000	1.000	1.000	10
n = 11	0	.8953	.8007	.7153	.6382	.5688	.5063	.4501	.3996	.3544	.3138	.1673	.0859	.0422	.0198	.0088	.0036	.0014	.0005	0
	1	.9948	.9805	.9587	.9308	.8981	.8618	.8228	.7819	.7399	.6974	.4922	.3221	.1971	.1130	.0606	.0302	.0139	.0059	1
	2	.9998	.9988	.9963	.9917	.9848	.9752	.9630	.9481	.9305	.9104	.7788	.6174	.4552	.3127	.2001	.1189	.0652	.0327	2
	3	1.000	1.000	.9998	.9993	.9984	.9970	.9947	.9915	.9871	.9815	.9306	.8389	.7133	.5696	.4256	.2963	.1911	.1133	3
	4			1.000	1.000	.9999	.9997	.9995	.9990	.9983	.9972	.9841	.9496	.8854	.7897	.6683	.5328	.3971	.2744	4
	5					1.000	1.000	1.000	.9999	.9998	.9997	.9973	.9883	.9657	.9218	.8513	.7535	.6331	.5000	5
	6								1.000	1.000	1.000	.9997	.9980	.9924	.9784	.9499	.9006	.8262	.7256	6
	7											1.000	.9998	.9988	.9957	.9878	.9707	.9390	.8867	7
	8												1.000	.9999	.9994	.9980	.9941	.9852	.9673	8
	9													1.000	1.000	.9998	.9993	.9978	.9941	9
	10															1.000	1.000	.9998	.9995	10
	11																	1.000	1.000	11
n = 12	0	.8864	.7847	.6938	.6127	.5404	.4759	.4186	.3677	.3225	.2824	.1422	.0687	.0317	.0138	.0057	.0022	.0008	.0002	0
	1	.9938	.9769	.9514	.9191	.8816	.8405	.7967	.7513	.7052	.6590	.4435	.2749	.1584	.0850	.0424	.0196	.0083	.0032	1
	2	.9998	.9985	.9952	.9893	.9804	.9684	.9532	.9348	.9134	.8891	.7358	.5583	.3907	.2528	.1513	.0834	.0421	.0193	2
	3	1.000	.9999	.9997	.9990	.9978	.9957	.9925	.9880	.9820	.9744	.9078	.7946	.6488	.4925	.3467	.2253	.1345	.0730	3
	4		1.000	1.000	.9999	.9998	.9996	.9991	.9984	.9973	.9957	.9761	.9274	.8424	.7237	.5833	.4382	.3044	.1938	4
	5				1.000	1.000	1.000	.9999	.9998	.9997	.9995	.9954	.9806	.9456	.8822	.7873	.6652	.5269	.3872	5
	6							1.000	1.000	1.000	.9999	.9993	.9961	.9857	.9614	.9154	.8418	.7393	.6128	6
	7										1.000	.9999	.9994	.9972	.9905	.9745	.9427	.8883	.8062	7
	8											1.000	.9999	.9996	.9983	.9944	.9847	.9644	.9270	8
	9												1.000	1.000	.9998	.9992	.9972	.9921	.9807	9
	10														1.000	.9999	.9997	.9989	.9968	10
	11															1.000	1.000	.9999	.9998	11
	12																	1.000	1.000	12
n = 13	0	.8775	.7690	.6730	.5882	.5133	.4474	.3893	.3383	.2935	.2542	.1209	.0550	.0238	.0097	.0037	.0013	.0004	.0001	0
	1	.9928	.9730	.9436	.9068	.8646	.8186	.7702	.7206	.6707	.6213	.3983	.2336	.1267	.0637	.0296	.0126	.0049	.0017	1
	2	.9997	.9980	.9938	.9865	.9755	.9608	.9422	.9201	.8946	.8661	.6920	.5017	.3326	.2025	.1132	.0579	.0269	.0112	2
	3	1.000	.9999	.9995	.9986	.9969	.9940	.9897	.9837	.9758	.9658	.8820	.7473	.5843	.4206	.2783	.1686	.0929	.0461	3
	4		1.000	1.000	.9999	.9997	.9993	.9987	.9976	.9959	.9935	.9658	.9009	.7940	.6543	.5005	.3530	.2279	.1334	4
	5				1.000	1.000	.9999	.9999	.9997	.9995	.9991	.9925	.9700	.9198	.8346	.7159	.5744	.4268	.2905	5
	6						1.000	1.000	1.000	.9999	.9999	.9987	.9930	.9757	.9376	.8705	.7712	.6437	.5000	6
	7									1.000	1.000	.9998	.9988	.9944	.9818	.9538	.9023	.8212	.7095	7
	8											1.000	.9998	.9990	.9960	.9874	.9679	.9302	.8666	8
	9												1.000	.9999	.9993	.9975	.9922	.9797	.9539	9
	10													1.000	.9999	.9997	.9987	.9959	.9888	10
	11														1.000	1.000	.9999	.9995	.9983	11
	12																1.000	1.000	.9999	12
	13																		1.000	13

2

It is important to note that the tables give a probability of *less* than or *equal* to a certain number of successes. So if you wanted to find the probability of *exactly* 4 successes then you would have to use the tables to find the probability of 4 successes or less by using the tables with $x = 4$. This would give the probability of 4, 3, 2, 1 or 0 successes. You could then use the tables to find the probability of 3 successes or less using $x = 3$. This would give the probability of 3, 2, 1 or 0. Thus by subtracting these two probabilities, the required probability of exactly 4 successes can be found using tables.

Hence, we have $P(X = 4) = P(X \leq 4) - P(X \leq 3)$,

Then for $n = 10$, $p = 0.3$,

$$P(X = 4) = 0.8497 - 0.6496$$
$$= 0.2001$$

You could still use the formula to find the probability and this would be a valid method if no indication of which method to use were given in the question.

Use of calculators instead of tables

Active Learning

Work through the text and work out how you would work out the values using your calculator.

The WJEC specification states that calculators must have certain functionalities, including the ability to compute summary statistics and access probabilities from standard statistical distributions. You must learn how to use the calculator because you may get questions which cannot easily be answered using the tables. Parts of the following examples have been answered using tables. See if you can obtain these answers using a calculator.

Examples

1 When cuttings of a certain plant are taken, the probability of each cutting rooting is 0.25 independently of all other cuttings.

Joshua takes 20 cuttings. Find the probability that at least 10 of the cuttings take root.

Answer

Here we will use the binomial distribution function table to work out the total probability of 9 or less taking root. Once found, we can take the answer from one to find the required probability.

1 Here we have $n = 20$, $p = 0.25$ and $x = 10$.

Using the tables we obtain a probability for $P(X \leq 9) = 0.9861$

Required probability, $P(X \geq 10) = 1 - 0.9861 = 0.0139$

---

2 The probability that a machine part fails in its first year is 0.05 independently of all other parts. In a batch of 20 randomly selected parts, find the probability that in the first year:

(a) exactly 1 part fails

(b) more than 2 parts fail.

Answer

2 (a) Using $P(X = x) = \binom{n}{x}p^x(1 - p)^{n-x}$,

with $x = 1$, $n = 20$ and $p = 0.05$ we obtain

Note that no method is specified in the question so for the answer you can use the binomial formula or use the tables.

$$P(X = 1) = \binom{20}{1}0.05^1(1 - 0.05)^{20-1}$$

$$= \binom{20}{1}0.05^1(0.95)^{19}$$

$$= 0.3773536$$

$$= 0.3774 \text{ (correct to 4 s.f.)}$$

Alternative method using tables with $n = 20$ and $p = 0.05$ we have

Note the slight difference in the answers between the calculated probability and the probability found using tables. This is caused by rounding off in the tables.

$$P(X = 1) = P(X \leq 1) - P(X \leq 0)$$

$$= 0.7358 - 0.3585$$

$$= 0.3773$$

(b) $P(X > 2) = 1 - P(X \leq 2)$

$= 1 - 0.9245$

$= 0.0755$

Tables have been used here but the binomial formula could also be used.

What to do if $p > 0.5$, as the tables do not show values above $p = 0.5$

It should be noted that in the tables for the binomial distribution the highest value of p considered is 0.50. When dealing with binomial distributions in which $p > 0.5$, we note that the probability of a failure $q = 1 - p < 0.5$. In this case, we consider the number of failures instead of the number of successes.

Examples

1 Given that X has the binomial distribution B (12, 0.6), find the values of:

(a) $P(X = 8)$

(b) $P(6 \leq X \leq 10)$.

Answer

1 Let $Y = 12 - X$

Then Y has the distribution B(12, 0.4)

X is the number of successes.
Y is the number of failures.

(a) $P(X = 8) = P(Y = 4) = P(Y \leq 4) - P(Y \leq 3)$

$= 0.4382 - 0.2253$

$= 0.2129$

(b) $P(6 \leq X \leq 10) = P(2 \leq Y \leq 6)$

$= P(Y \leq 6) - P(Y \leq 1)$

$= 0.8418 - 0.0196$

$= 0.8222$

---

2 The probability that a randomly chosen daffodil bulb will produce a flower is 0.8. If 20 such bulbs are planted, find the probabilities that:

(a) exactly 12 of them will produce flowers

(b) fewer than 8 will produce flowers.

Answer

2 Let X be the number of bulbs that produce flowers, so that X is distributed as B(20, 0.8).

Since $p = 0.8 > 0.5$ we consider $Y = 20 - X$ and note that Y is distributed as B(20, 0.2).

X is the number of successes.
Y is the number of failures.

(a) $P(X = 12) = P(Y = 8)$

$= P(Y \leq 8) - P(Y \leq 7)$

$= 0.9900 - 0.9679$

$= 0.0221$

(b) $P(X < 8) = P(X \leq 7) = P(Y \geq 13)$

$P(Y \geq 13) = 1 - P(Y \leq 12)$

$= 1 - 1$

$= 0$

3 (a) A series of trials is carried out, each resulting in either success or failure. State two conditions that have to be satisfied in order for the total number of successes to be modelled by the binomial distribution. [2]

(b) Each time Ann shoots an arrow at a target, she hits it with probability 0·4. She shoots 20 arrows at the target. Determine the probability that she hits it:

(i) exactly 8 times

(ii) between 6 and 10 times (both inclusive). [5]

Answer

3 (a) The two conditions are:

- independent trials,
- trials where there is a constant probability of success.

(b) (i) $P(X = x) = \binom{n}{x}p^x(1 - p)^{n-x}$

Now $p = 0.4$, $n = 20$, $x = 8$, so we have

$$P(X = 8) = \binom{20}{8}0.4^8(1 - 0.4)^{20-8}$$

$$P(X = 8) = \binom{20}{8}0.4^8(0.6)^{12}$$
$$= 0.1797 \text{ (correct to 4 s.f.)}$$

(ii) $P(6 \le X \le 10) = P(X \le 10) - P(X \le 5)$
$$= 0.8725 - 0.1256$$
$$= 0.7469 \text{ (correct to 4 s.f.)}$$

BOOST Grade

Questions on the conditions for which a certain modelling formula can be used are frequent. Make sure you remember the conditions for each formula you use.

The formula is obtained from the formula booklet.

$P(X = 8) = P(X \le 8) - P(X \le 7)$
$= 0.5956 - 0.4159$
$= 0.1797$ (to 4 s.f.)

4 Wine glasses are packed in boxes, each containing 20 glasses. Each glass has a probability of 0·05 of being broken in transit, independently of all other glasses.

Let X denote the number of glasses in a box broken in transit.

(a) State the distribution of X.

(b) **Without** the use of tables, calculate $P(X = 1)$.

(c) **Using tables**, determine the value of $P(X \ge 3)$. [5]

Answer

4 (a) The distribution is B(20, 0.05)

(b) $P(X = 1) = \binom{20}{1}0.05^1(1 - 0.05)^{20-1}$

$$= \binom{20}{1}0.05^1(0.95)^{19}$$
$$= 0.377 \text{ (correct to 3 s.f.)}$$

Insert $n = 20$ and $p = 0.05$ into $B(n, p)$ to give $B(20, 0.05)$.

(c) $P(X \geq 3) = 1 - P(X \leq 2) = 1 - 0.9245$

$= 0.0755$

Use the formula $P(X = x) = \binom{n}{x} p^x (1 - p)^{n-x}$ obtained from the formula booklet.

---

5 When seeds of a certain variety of flower are planted, the probability of each seed germinating is 0·8, independently of all other seeds.

(a) David plants 20 of these seeds. Find the probability that:

(i) exactly 15 seeds germinate

(ii) at least 15 seeds germinate. [6]

(b) Beti plants n of these seeds and she correctly calculates that the probability that they all germinate is 0.10737, correct to five decimal places. Find the value of n. [3]

Answer

5 (a) (i) Distribution is B(20, 0.8)

$P(X = 15) = \binom{20}{15} 0.8^{15} (1 - 0.8)^5$

$= 0.1746$

Or Let Y = no. of seeds failing to germinate.

Y is distributed as B(20, 0.2)

$P(X = 15) = P(Y = 5)$

$= P(Y \leq 5) - P(Y \leq 4)$

$= 0.8042 - 0.6296$

$= 0.1746$

(ii) Let the number of seeds failing to germinate = Y.

Y is distributed as B(20, 0.2)

$P(X \geq 15) = P(Y \leq 5) = 0.8042$

Use the binomial cumulative distribution function table with $n = 20$, $p = 0.2$ and $x = 5$. Read off the probability, which is 0.8042.

(b) Distribution is B(n, 0.8)

$P(X = n) = \binom{n}{n} 0.8^n (1 - 0.8)^0$

$0.10737 = 0.8^n$

Taking $\log_e$ of both sides, we obtain

$\log_e 0.10737 = \log_e 0.8^n$

so $\log_e 0.10737 = n \log_e 0.8$

$$n = \frac{\log_e 0.10737}{\log_e 0.8}$$

Hence, $n = 10$

Note that you could also take logarithms of both sides to base 10 to solve this equation.

Note that as n is the number of seeds planted, n has to be an integer.

4.2 The Poisson distribution as a model

The Poisson distribution is a discrete probability distribution and is used to model the number of events occurring randomly within a given interval of time and space.

In a particular interval, the probability of an event X occurring x number of times is given by the following formula

Note that this formula is given in the formula booklet.

$$P(X = x) = e^{-\lambda}\frac{\lambda^x}{x!}$$

where $\lambda = \mu = E(X)$ and $x = 0, 1, 2, 3, 4, \ldots$

If the probabilities of X are distributed in this way, we write

$$X \sim Po(\lambda)$$

Note that the symbol $\sim$ means 'is distributed'.

Lambda, λ, is the parameter of the distribution. Sometimes you will see the mean m replacing lambda as the parameter. We say X is distributed with a Poisson with parameter λ.

Example

1 Use the random variable $X \sim Po\ (1.4)$ to determine:

(a) $P(X = 2)$

(b) $P(X \geq 1)$

(c) $P(2 < X \leq 4)$

Answer

The formula, $P(X = x) = e^{-\lambda}\frac{\lambda^x}{x!}$ is looked up in the formula booklet and used here.

1 (a) $P(X = 2) = e^{-1.4}\frac{1.4^2}{2!}$

$= 0.2417$

P($X \geq 1$) means the probability of x being 1, 2, 3, ... Note that it does not include the probability of x being 0. Hence we can subtract the probability of P($X = 0$) from 1.

(b) $P(X \geq 1) = 1 - P(X = 0)$

$= 1 - \left(e^{-1.4}\frac{1.4^0}{0!}\right)$

$= 1 - 0.2466$

$= 0.7534$

Note that $1.4^0 = 1$ and also that $0! = 1$.

(c) $P(2 < X \leq 4) = P(X = 3) + P(X = 4)$

$= e^{-1.4}\frac{1.4^3}{3!} + e^{-1.4}\frac{1.4^4}{4!}$

$= 0.1128 + 0.0395$

$= 0.152$ (correct to 3 decimal places)

Using tables we obtain the following results

(a) $P(X = 2) = P(X \leq 2) - P(X \leq 1) = 0.8335 - 0.5918 = 0.2417$

(b) $P(X \geq 1) = 1 - P(X = 0) = 1 - 0.2466 = 0.7534$

(c) $P(2 < X \leq 4) = P(3 \leq X \leq 4) = P(X \leq 4) - P(X \leq 2) = 0.9857 - 0.8335$

$= 0.1522$

BOOST Grade

Remember to read inequalities very carefully. Many students fail to do this. Here many students will, incorrectly, also include P($X = 2$).

When to use the Poisson distribution

Both the binomial and Poisson distributions are discrete probability distributions.

In general we use the Poisson distribution in the following circumstances:

- If n is large (usually > 50) **and**
- p is small (usually <0.1).

Using Poisson distribution function tables to work out $P(X \le x)$

Poisson distribution function tables can be used to work out $P(X \le x)$. For example, if you wanted to find $P(X \le 3)$ for a distribution Po(0.8) you would look up the intersection of the row where $x = 3$ with the column where λ or $m = 0.8$ to give the required probability of 0.9909.

Examples

1 Given that 5% of pupils in a school are left-handed, use the Poisson distribution to estimate the probability that a random sample of 100 pupils in the school contains two or more left-handed pupils.

Answer

1 $\lambda = np = 100 \times 0.05 = 5$

X is approximately distributed as Po(5)

$$P(X \ge 2) = 1 - P(X \le 1)$$
$$= 1 - 0.0404$$
$$= 0.9596$$

---

2 The random variable X has the binomial distribution B(300, 0.012). Use a Poisson approximation to find an approximate value for the probability that X is less than 3.

Answer

2 Mean $= np = 300 \times 0.012 = 3.6$

X is approximately distributed as Po(3.6) (i.e. $X \sim$ Po(3.6))

$$P(X < 3) = P(X \le 2)$$
$$= 0.3027$$

Compare B(300, 0.012) with B(n, p) gives $n = 300$ and $p = 0.012$.

Note that this means the parameter $\lambda = 3.6$.

---

3 Cars arrive at a petrol station in such a way that the number arriving during an interval of length t minutes has a Poisson distribution with mean $0.2t$.

(a) Find the probability that:

(i) exactly ten cars arrive between 9 a.m. and 10 a.m.,

(ii) more than five cars arrive between 11 a.m. and 11.30 a.m. [6]

(b) The probability that no cars arrive during an interval of length t minutes is equal to 0·03. Without the use of tables, find the value of t. [4]

Answer

3 (a) (i) Mean $\lambda = 0.2t$ and over 60 min (i.e. 1 hour) $\lambda = 0.2 \times 60 = 12$

X is distributed as Po(12)

$$P(X = 10) = e^{-12}\frac{12^{10}}{10!} = 0.1048 \text{ (correct to 4 s.f.)}$$

$P(X = 10)$
$= P(X \leq 10) - P(X \leq 9)$
$= 0.3472 - 0.2424$
$= 0.1048$

(ii) Mean $\lambda = 0.2t$ and over 30 min (i.e. 0.5 hour) $\lambda = 0.2 \times 30 = 6$

X is Po(6)

$$P(X > 5) = 1 - P(X \leq 5)$$
$$= 1 - 0.4457$$
$$= 0.5543 \text{ (correct to 4 s.f.)}$$

Note that $P(X \leq 5)$ is found by using tables.

(b) Mean $\lambda = 0.2t$

$$P(X = 0) = e^{-0.2t}\frac{(0.2t)^0}{0!}$$

$$P(X = 0) = e^{-0.2t}$$

Note that $(0.2t)^0 = 1$ and also that $0! = 1$

Now, $P(X = 0) = 0.03$

Hence, $e^{-0.2t} = 0.03$

Taking $\log_e$ of both sides, we obtain

$$-0.2t = \log_e 0.03$$

Solving, gives $t = 17.5$ minutes (correct to one decimal place).

Note that the solving of equations by the use of logarithms was covered in Unit 1. Instead of taking $\log_e$ of both sides you could have used $\log_{10}$.

---

4 The number of parcels per day arriving at a school has a Poisson distribution with a mean of 2.5, independently of all other days.

(a) Without the use of tables, find the probability that on a randomly selected day:

(i) no parcels arrive

(ii) either 4 or 5 parcels arrive. [5]

(b) Over a 3-day period, calculate the probability that:

(i) no parcels arrive

(ii) the first parcel that arrives, arrives on the third day. [5]

Answer

4 (a) (i) Using $P(X = x) = e^{-\lambda}\frac{\lambda^x}{x!}$, we obtain

The equation used here is $P(X = x) = e^{-\lambda}\frac{\lambda^x}{x!}$ which is obtained from the formula booklet.

$$P(X = 0) = e^{-2.5}\frac{2.5^0}{0!}$$
$$= 0.0821$$

Note that $2.5^0 = 1$ and also that $0! = 1$.

(ii) Using $P(X = 4 \text{ or } 5) = e^{-2.5}\frac{2.5^4}{4!} + e^{-2.5}\frac{2.5^5}{5!}$

$$= 0.1336 + 0.0668$$
$$= 0.200 \text{ (correct to 3 decimal places)}$$

The probability of 4 parcels arriving is found and added to the probability of 5 parcels arriving.

(b) (i) P(no parcels arriving over 3 days) = 0.0821 × 0.0821 × 0.0821

$= 0.0821^3$

= 0.00055

(ii) There must be no parcels on the first day and the second day.

P(no parcels arriving until the third day) = 0.0821 × 0.0821 × (1 − 0.0821)
= 0.0062 (correct to 4 decimal places)

Notice that the probabilities are multiplied together. You can only do this for independent events (i.e. where the probabilities stay constant).

Using the distribution tables backwards

In the examples involving use of the probability distribution tables, you have found the probability. You can also use the tables backwards where you are given the probability and have to find a missing value (usually x).

Suppose at a charity garden fete there is a game where a coin is tossed 10 times and the person tossing the coin has to toss x heads or more to win a prize. It has been decided that the probability of winning a prize must be less than 0.1

Find the minimum number of heads needed to win a prize.

Let X be the number of heads tossed.

$\text{P}(X \geq x) < 0.1$

$\text{P}(X < x) > 1 - 0.1 > 0.9$

We need to find $\text{P}(X < x)$ as the tables give values of at least x successes.

The situation can be modelled as B(10, 0.5)

Here $p = 0.5$, $n = 10$ and the probability in the main body of the table is 0.1.

Looking up x in the table gives

$\text{P}(X \leq 6) = 0.8281$ so $\text{P}(X > 6) = 1 - 0.8281 = 0.1719$ (this is higher than 0.1)

$\text{P}(X \leq 7) = 0.9453$ so $\text{P}(X > 7) = 1 - 0.9453 = 0.0547$ (this is lower than 0.1)

Hence the number of heads needed to win a prize should be set at 7 or more.

Using the binomial formula backwards

In most questions you will know n and X or a range of values of X and you have to find the probability. In this example you are given the probability and X and are asked to find n.

Examples

1 In autumn Chloe plants snowdrop bulbs. She knows not all of them will grow as some bulbs are eaten by mice. The probability that each bulb will grow in spring is 0.7, independent of other bulbs growing.

Chloe plants n bulbs and she correctly calculates that the probability that they all grow in spring is 0.001628 correct to four significant figures. Find the value of n.

Answer

1 $P(X = x) = \binom{n}{x} p^x (1 - p)^{n-x}$

As all grow $x = n$ and $P(X = n) = 0.001628$

$$P(X = n) = \binom{n}{n} 0.7^n (1 - 0.7)^{n-n}$$

$$= \binom{n}{n} 0.7^n (1 - 0.7)^0$$

$$= (0.7)^n$$

$$P(X = n) = 0.001628$$

Note that $\binom{n}{n} = 1$ and $(1 - 0.7)^0 = 1$

So $(0.7)^n = 0.001628$

Taking $\log_e$ of both sides

$$\log_e 0.7^n = \log_e 0.001628$$

$$n \log_e 0.7 = \log_e 0.001628$$

$$n = \frac{\log_e 0.001628}{\log_e 0.7}$$

$$= 18$$

---

2 In a multiple choice test, students have to choose an answer from A, B, C or D. The exam board would like the probability of a student passing the test by guessing the answers to be 0.1 or less. In a test consisting of 20 questions with one mark for each correct answer, what should the pass mark be so there is less than 0.1 probability of passing the test by simply guessing the answer?

Answer

2 The probability of guessing the correct answer, p, is 0.25 (i.e. 1 out of 4 choices)

The student simply guessing the answers to the 20 questions can be modelled as the binomial distribution B(20, 0.25).

Let x be the mark at which the probability of obtaining x or more is less than 0.1.

Hence $P(X \geq x) < 0.1$

As the tables refer to probabilities of at most x successes we need to find $P(X < x)$.

Now $P(X < x) > 1 - 0.1 = 0.9$

The binomial distribution function tables are now used to determine values for x which give a probability near to, or exactly equal to, 0.9.

So we need to read off $n = 20$ and $p = 0.25$ to find the appropriate value of x.

$$P(x \leq 7) = 0.8982 \text{ and } P(x \leq 8) = 0.9591$$

Probability of obtaining more than 7 correct answers through guessing is

$$1 - 0.8982 = 0.1018, \text{ which is larger than } 0.1$$

Probability of obtaining more than 8 correct answers through guessing is

$1 - 0.9591 = 0.0409$, which is smaller than 0.1.

This means the pass mark needs to be set at more than 8 marks.

Hence, the pass mark needs to be a minimum of 9 marks so that the probability of passing by guessing is 0.1.

---

3 A newsagent sells the *Daily Bugle* newspaper. You may assume that the daily demand for this newspaper has a Poisson distribution with mean 15. The newsagent begins each day with 20 copies of the newspaper.

(a) Calculate the probability that, on a randomly chosen day, the newsagent sells:

(i) 12 copies of the newspaper,

(ii) all 20 copies of the newspaper. [4]

(b) Determine the minimum number of copies of the *Daily Bugle* that the newsagent should buy each day in order to satisfy the demand with a probability of at least 0·99. [2]

Answer

3 (a) (i) $P(X = x) = e^{-\lambda}\frac{\lambda^x}{x!}$

$P(X = 12) = e^{-15}\frac{(15)^{12}}{12!}$

$= 0.083$

(ii) $P(X \geq 20) = 1 - P(X \leq 19)$

To determine $P(X \leq 19)$ use the table for the Poisson distribution function and look up the probability for $x = 19$ and $m = 15$.

From the table, $P(X \leq 19) = 0.8752$

$P(X \geq 20) = 1 - 0.8752$

$= 0.1248$

(b) The Poisson distribution table is used with mean, $m = 15$ and $P(X \leq x) = 0.99$ to find the value of x.

$x = 25$ gives a probability of 0.9938 which is the nearest value to 0.99.

Number of copies = 25

It is always very important to read and understand the question. It is the demand that follows the Poisson distribution, so it is possible that the demand can be greater than the 20 papers that are for sale. We need to find the probability of the demand for papers being 20 or more. Hence we need to find $P(X \geq 20)$.

$P(X \leq x) = 0.99$ is found in the body of the table.

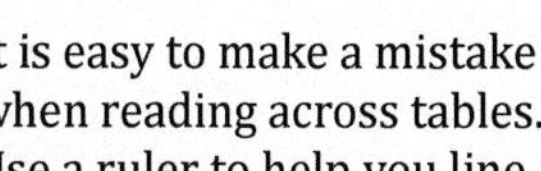

It is easy to make a mistake when reading across tables. Use a ruler to help you line the numbers up.

4.3 The discrete uniform distribution as a model

A discrete uniform distribution is a distribution where all the outcomes are equally likely. For example, a fair dice has an equal probability of landing on one of the numbers 1 to 6 when thrown.

The discrete uniform distribution for this can be shown using this table:

x	1	2	3	4	5	6
$P(X = x)$	$\frac{1}{6}$	$\frac{1}{6}$	$\frac{1}{6}$	$\frac{1}{6}$	$\frac{1}{6}$	$\frac{1}{6}$

The distribution can alternatively be shown as a diagram:

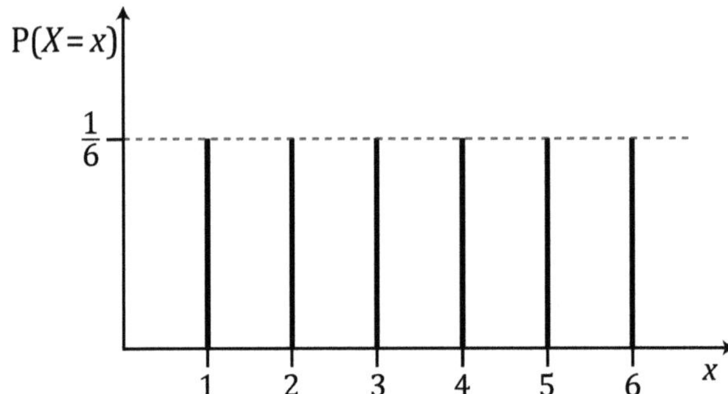

A discrete random variable has a discrete uniform distribution when each value of the random variable is equally likely, and values are uniformly distributed throughout some interval. In the case above the interval is from 1 to 6.

If X is a discrete variable and is uniformly distributed on the set $\{1,2,3,4...,N\}$ then the following formula applies:

$$P(x) = \frac{1}{N}$$

Example

1 A hexagonal spinner numbered from 1 to 6 is spun and the number obtained X is recorded. This process is repeated a set number of times.

(a) Explain what conditions must be satisfied if the number obtained X is to be modelled as a discrete uniform distribution.

(b) Find $P(1 \leq X < 4)$

Answer

1 (a) The numbers obtained must be random (i.e. no bias) so that each number 1 to 6 has an equal probability of occurring $\left(\text{i.e. } \frac{1}{6}\right)$.

(b) $P(1 \leq X < 4)$ means the probability of a spin giving a 1, 2 or 3.

$$P(1 \leq X < 4) = \frac{1}{6} + \frac{1}{6} + \frac{1}{6} = \frac{1}{2}$$

4.4 Selecting an appropriate probability distribution

In examination questions, you will not necessarily be told which probability distribution to use for your answer. Instead you have to decide which one to use. Here is some guidance you can use to help decide.

The discrete uniform distribution

When each value of a discrete random variable X has the same probability of occurring, the distribution is called a discrete uniform distribution.

The binomial distribution

The binomial distribution can be used to work out probabilities when the following conditions are satisfied:

1. There must be a fixed number of trials (as denoted in the formula as n).
2. There must be an exact probability of the event occurring.
3. Each trial must be independent. (i.e. each trial should not have any effect on any other trial).
4. There must be only two outcomes (i.e. success and failure).
5. The probability of success must be constant (i.e. there is a constant value for p the letter used for the probability of success).

The Poisson distribution

The Poisson distribution is used to model rare events that occur over a period of space or time. Poisson does not take into account whether a particular event will or won't occur.

You can usually get an indication of when to use the Poisson distribution if the words 'average' or 'mean' are mentioned in the question, as this means you have a value for λ.

You would use Poisson if the following conditions are satisfied:

1. There are an infinite number of trials.
2. You know the mean/average rate for the events happening.
3. Each trial must be independent.
4. There is success or failure.

The Poisson distribution is therefore used when random events happen one at a time with a constant rate.

Examples

In some examination questions you will not be told in the question which distribution to use.

1. Think about the following situations and decide whether they are best modelled by a binomial distribution or a Poisson distribution. You should explain your answer.

(a) A customer service department receives an average of 15 complaints per hour. What is the probability that there are at most 10 complaints in a certain period of one hour?

(b) The demand for rental cars from a car hire firm has a mean of 8 per day. Find the probability that on a randomly chosen day, the demand is for 8 or more cars.

(c) A bag contains 6 red and 4 black balls. One ball is taken out, its colour noted and then returned to the bag. The process is repeated until 5 balls' colours have been recorded. What is the probability that all 5 balls are red?

(d) 5% of a batch of components are faulty. Find the probability that in a randomly selected batch of 20 components, 4 components are faulty.

(e) A test containing multiple choice questions with a choice of answers A, B, C or D consists of 20 questions. The question setter needs to be sure that the chance of passing the test simply by guessing the answers is less than 0.05. How many marks should the pass mark be set at for the probability of passing by guessing to be less than 0.05?

(f) Daisies are growing randomly in a meadow. There is an average of six daisy plants per square metre of meadow. Find the probability that a randomly picked area of 0.5m^2 of the meadow contains no daisies.

(g) A salesman in a car showroom sells on average 12 cars per week. What is the probability that he sells less than 5 cars in a randomly selected week?

(h) A one-mile stretch of road contains an average of 3 potholes. Find the probability that in a half-mile stretch of the same road there will be no potholes.

(i) It is known that a packet of seeds has a probability of 0.95 of all the seeds germinating. Find the probability that in a packet of 50 seeds at most 5 seeds will not germinate.

(j) In an accident and emergency department it is known that the average wait to be seen by a doctor is 2 hours. Find the probability of a patient turning up on a random day and waiting less than 1 hour.

Answer

1 (a) The clue is in the word 'average' in the question. As the average rate is known this is solved using Poisson.

(b) Here the word 'mean' is used and a rate of 8 per day is given so this can be solved using Poisson.

(c) Here there are a fixed number of trials (i.e. 5) with an exact probability of success $\left(\text{i.e. chance of a red ball} = \frac{6}{10}\ or\ \frac{3}{5}\right)$. This can be solved using binomial.

(d) The exact probability of success (i.e. a faulty component) is known and the number of trials is known so this can be solved using binomial.

(e) There is a known probability of selecting the correct answer to a question by guessing $\left(\text{i.e. } \frac{1}{4}\right)$. This will be the probability of success.

The total number of trials is known (i.e. 20) so this can be solved using binomial.

(f) Here you are given a rate of 6 per square metre so this indicates Poisson. Also the word average in the second sentence of the question indicates Poisson.

(g) Here an average rate is given so this is Poisson.

(h) Notice the word average is used and also there is a rate mentioned (i.e. 3 potholes per mile) so this is Poisson.

(i) Notice the accurate probability of 0.95 and also the number of seeds in the pack. This means both p and n are known so this is binomial.

(j) The average is given and there is no exact probability, so this is Poisson.

---

2 Cars arrive at random at a toll bridge at a mean rate of 15 per hour.

(a) Explain briefly why the Poisson distribution could be used to model the number of cars arriving in a particular time interval. [1]

(b) Phil stands at the bridge for 20 minutes. Determine the probability that he sees exactly 6 cars arrive. [3]

(c) Using the statistical tables provided, find the time interval (in minutes) for which the probability of more than 10 cars arriving is approximately 0.3. [3]

Answer

2 (a) The Poisson distribution can be used to model randomly occurring events in a certain period of time when they occur independently and at a mean rate.

(b) There are15 arrivals per hour so in 20 minutes ($\frac{1}{3}$ of an hour) there would be a mean rate of arrivals of 5 cars.

Mean rate of arrivals $\lambda = 5$

The Poisson formula is looked up: $P(X = x) = e^{-\lambda}\frac{\lambda^x}{x!}$

This is in the formula booklet so you don't need to remember it.

$$P(X = 6) = e^{-5}\frac{5^6}{6!}$$

$$= 0.1462 \text{ (4 s.f.)}$$

(c) $P(X > 10) = 1 - P(X \leq 10)$

Now $P(X > 10) \approx 0.3$ (you are told this in the question)

Hence $0.3 \approx 1 - P(X \leq 10)$ so $P(X \leq 10) \approx 0.7$

We now use the Poisson distribution function tables to determine the value of the mean for $X \leq 10$ with a probability of approximately 0.7.

The nearest we can find to 0.7 is 0.7060 and this gives a mean of 9.

The mean rate of arrival is 15 per hour which is 1 every 4 minutes.

So time at the bridge = 9 × 4 = 36 minutes

Test yourself

1. It is known that 25% of the bulbs in a box produce yellow flowers. A customer buys 20 of these bulbs. Find the probability that:
   (a) exactly 4 bulbs produce yellow flowers
   (b) fewer than 8 bulbs produce yellow flowers.

2. The number of items of junk mail arriving by post each day at a house can be modelled by a Poisson distribution with mean 3.4 .
   (a) Without using tables, calculate:
      (i) $P(X = 4)$
      (ii) $P(X \leq 2)$.
   (b) Using tables, determine $P(4 \leq X \leq 7)$.

3. Each time a darts player throws a dart at the bulls-eye they hit the bulls-eye with a probability 0.08. The darts player throws 100 darts at the bulls-eye. Use a Poisson approximation to find the probability that she hits the bulls-eye fewer than 5 times.

4. On a turtle farm, turtles are bred and hatched from eggs under controlled conditions.
   (a) The probability of producing a female turtle from an egg is 0.4 under the controlled conditions. The probability of producing a female from an egg is independent of other eggs hatching to produce female turtles. When 20 eggs are kept under the controlled conditions, find the probability that:
      (i) exactly 10 female turtles are produced
      (ii) more than 7 female turtles are produced. [5]
   (b) During the hatching process, the probability that an egg fails to hatch is 0.05. When 300 eggs are kept under the controlled conditions, use the Poisson approximation to find the probability that the number of eggs failing to hatch is less than 10. [3]

5. (a) A factory manufactures cups. The manager knows from past experience that 5% of the cups produced are defective. Given a random sample of 50 of these cups, determine the probability that the number of defective cups in this sample is:
      (i) exactly 2
      (ii) between 3 and 8 (both inclusive). [6]
   (b) The factory also manufactures plates. The manager knows that 1·5% of the plates produced are defective. A random sample of 250 plates is taken.
      (i) Explain why the Poisson distribution can be used as an approximation to the binomial distribution, to model the number of plates that are defective.
      (ii) Use an appropriate Poisson distribution to find, approximately, the probability that the sample of plates contains exactly 4 defective plates. [5]

6 (a) The random variable X has the binomial distribution $B(20, 0{\cdot}2)$.
   (i) Without the use of tables, calculate $P(X = 6)$,
   (ii) Determine $P(2 \leq X \leq 8)$. [5]

(b) The random variable Y has the binomial distribution $B(200, 0{\cdot}0123)$. Use the Poisson distribution to determine the approximate value of $P(Y = 3)$. [3]

7 (a) When a certain type of seed is planted, there is a probability of 0.7 that it produces red flowers. A gardener plants 20 of these seeds. Calculate the probability that:
   (i) exactly 15 seeds produce red flowers
   (ii) more than 12 seeds produce red flowers. [6]

(b) When a different type of seed is planted, there is a probability of 0.09 that it produces white flowers. The gardener plants 150 of these seeds. Use an appropriate Poisson distribution to determine, approximately, the probability that exactly 10 seeds produce white flowers. [3]

Summary

The binomial distribution

For a fixed number of trials, n, each with a probability p of occurring, the probability of a number x of successes is given by the formula:

$$P(X = x) = \binom{n}{x} p^x (1 - p)^{n-x}$$

Check you know the following facts:

Conditions for using the binomial distribution

The conditions for using the binomial distribution are:

- Independent trials (i.e where the probability of one event does not depend on another event).
- Trials where there is a constant probability of success.
- A fixed number of trials.
- Where there is only success or failure.

The Poisson distribution

In a particular interval, the probability of an event X occurring x times is given by the following formula:

$$P(X = x) = e^{-\lambda} \frac{\lambda^x}{x!}$$

$$\text{where } \lambda = \mu = E(X) \text{ and } x = 0, 1, 2, 3, 4, \ldots$$

The mean of the Poisson distribution

If X is $Po(\lambda)$ then mean, $\mu = \lambda$

When to use the Poisson distribution

In general we use the Poisson distribution in the following circumstances:

- If n is large (usually > 50) ***and***
- p is small (usually <0.1).

The discrete uniform distribution

This is a distribution where all the outcomes are equally likely.

So if there are N possible outcomes, the probability of a particular outcome $= \frac{1}{N}$.

5 Statistical hypothesis testing

Introduction

The purpose of collecting data is to analyse it and ask questions about it and see if certain hypotheses about a parameter (e.g. probability, mean, etc.) of the population are true or false. For example, there might be a statement made by the head teacher of a school that '90% of all pupils in the school think the teaching in the school is excellent'. A sample of students could be asked if they thought the teaching was excellent and the results could be tested to see if this statement is justified or not. Hypothesis testing uses such data to test to see if a hypothesis should be accepted or rejected in favour of an alternative hypothesis.

This topic covers the following:

- 5.1 Understanding and applying the language of hypothesis testing
- 5.2 Finding critical values and critical regions and using significance levels
- 5.3 One- and two-tailed tests
- 5.4 Interpreting and calculating type I and type II errors
- 5.5 Conducting a hypothesis using *p*-values

5.1 Understanding and applying the language of hypothesis testing

There are quite a few specialist terms used in hypothesis testing and you need to remember what each of them means.

What is a parameter?

Remember that it is usually too difficult and sometimes impossible to collect data about the entire population. Instead data is collected about a smaller representative sample. The data can be collected by performing a survey or an experiment.

Do you remember the terms **population** and **sample**? A **population** is everything under consideration (e.g. students in a school, people eligible to vote in an election, red squirrels in a wood, etc.) and a **sample** is a smaller representative number about which data can be collected. If the sample has been collected correctly, any statistics, conclusions, etc., obtained using the sample should apply to the much larger population.

A **parameter** is a certain characteristic that can be used to describe a population. The parameter used in this topic will be the probability p, which is the probability of success in a single trial. For example, if a fair coin was tossed 30 times, the probability p would be 0.5 as this is the probability of obtaining a head (or alternatively a tail) in a single trial (i.e. a toss of the coin).

What is hypothesis testing?

A **hypothesis** is an assumption or claim about a certain value of a population parameter. In this topic the population parameter used will be p.

The assumption may be true or false and hypothesis testing is a statistical procedure that allows us to determine whether a hypothesis should be accepted or rejected.

There are two types of statistical hypotheses:

Null hypothesis
The null hypothesis is given the symbol $\mathbf{H}_0$ and is the currently accepted value for the parameter. For example, if we were checking to see if a coin was biased in favour of heads, the null hypothesis would be that it wasn't biased and the probability p of obtaining a head was 0.5.

The null hypothesis is the hypothesis where nothing has changed or nothing is biased. Data collected from the population in the form of a sample may cause you to reject the null hypothesis. For example, on a certain number of tosses of the coin you might get a larger number of heads than you might expect which causes you think the coin might be biased in favour of heads.

The alternative hypothesis is always the opposite to the null hypothesis.

Alternative hypothesis
The alternative hypothesis is given the symbol $\mathbf{H}_1$, and is considered to be the research hypothesis and this involves a claim that needs to be tested. The alternative hypothesis is where something is changed or is biased.

Note that the null hypothesis is the status quo (i.e. nothing has changed so the coin is not biased). The alternative hypothesis is that something has changed (i.e. the coin is biased).

In the coin tossing example the alternative hypothesis is that the coin is biased in favour of heads, so the probability of obtaining a head is greater than 0.5. Hence the alternative hypothesis is $p > 0.5$.

These two hypotheses can be expressed as follows:

$$\mathbf{H}_0 : p = 0.5$$

$$\mathbf{H}_1 : p > 0.5$$

where $\mathbf{H}_0$ is the null hypothesis and $\mathbf{H}_1$ is the alternative hypothesis.

As the alternative hypothesis has only one inequality to consider (i.e. $\mathbf{H}_1: p > 0.5$) we are investigating the probability in one direction only and this is called a one-tailed test. This will be looked at in more detail later.

BOOST Grade

You are often asked to give the null hypothesis and the alternative hypothesis for a situation in exam questions so make sure you stick to the format given above.

The test statistic

In order to test a hypothesis, you need a test statistic which is a statistic collected from the sample data to see whether the null hypothesis should be rejected or not.

Suppose an experiment is conducted to see how many times, X, the coin landed on heads when it is tossed 50 times.

The test statistic here is X, the number of heads in 50 tosses.

What if we think the coin might be biased but we don't know whether it is biased in favour of heads or tails?

Suppose we suspect the coin is biased but we do not know whether it is biased in favour of heads or tails.

If the coin is unbiased, the probability of obtaining a head, p is 0.5 so the null hypothesis is $p = 0.5$ which can be written using symbols as $\mathbf{H}_0 : p = 0.5$.

If the coin is biased, the probability of obtaining a head, p is not equal to 0.5 so the alternative hypothesis is $p \neq 0.5$ which is $\mathbf{H}_1 : p \neq 0.5$. Another way of considering the alternative hypothesis is to say that if p is not equal to 0.5 it must be either greater than 0.5 or less than 0.5. Hence, $\mathbf{H}_1 : p \neq 0.5$ can be written as the two tests:

$$\mathbf{H}_1 : p > 0.5 \quad \text{or} \quad p < 0.5.$$

As there are now two probabilities to consider, it is called a **two-tailed test**. This will be looked at in more detail later.

Examples

1 A four-sided spinner, with sides numbered 1 to 4, is used in a game. Jack suspects that the spinner is biased in favour of the number one. He wants to perform an experiment to test if the spinner is biased towards one. He spins the spinner 10 times and records the number of times it lands on number one.

(a) Write a statement for the null hypothesis that can be tested.

(b) Write a statement for the alternative hypothesis.

(c) Describe a test statistic that Jack could use.

Answer

1 (a) $\mathbf{H}_0 : p = \frac{1}{4}$

(b) $\mathbf{H}_1 : p > \frac{1}{4}$

(c) The test statistic is X, the number of times the spinner lands on number one in ten spins.

$\mathbf{H}_0$ uses the accepted view which is the probability that the spinner is unbiased, so the probability of it landing on 1 is $\frac{1}{4}$.

2 It is claimed that a six-sided dice is biased towards the number three. The dice is tossed 300 times and the number of times it lands on three is recorded.

(a) Write a statement for the null hypothesis that can be tested.

(b) Write a statement for the alternative hypothesis.

(c) Describe a test statistic that could be used to test the claim.

Answer

Note that for the null hypothesis we assume that the dice is not biased, so the probability of obtaining a 3 is $\frac{1}{6}$.

2 (a) $\mathbf{H}_0 : p = \frac{1}{6}$

(b) $\mathbf{H}_1 : p > \frac{1}{6}$

(c) The test statistic is X, the number of times the dice lands on number 3 in three hundred tosses.

---

3 A general election is to take place. Candidate A says she has 47% of the vote but the local newspaper thinks she is overestimating her support. The newspaper conducts a survey using a sample of 500 people eligible to vote, asking who they would vote for.

(a) Write a statement for the null hypothesis that can be tested.

(b) Write a statement for the alternative hypothesis.

(c) Describe a test statistic that could be used to test the claim made by candidate A.

Answer

3 (a) $\mathbf{H}_0 : p = 0.47$

(b) $\mathbf{H}_1 : p < 0.47$

(c) The test statistic is X, the number of people who said they would vote for candidate A out of the sample.

---

4 John thinks a coin is biased so he conducts an experiment using 100 tosses of the coin and he records the number of times it landed on heads and tails.

(a) Write down a suitable null hypothesis he could use to test the coin.

(b) Write down an alternative hypothesis he could use to test the coin.

(c) Describe the test statistic that could be used to check for bias.

Answer

Note that the alternative hypothesis uses both $p > 0.5$ and $p < 0.5$ which means the probability is being tested in both directions. When this happens this is called a two-tailed test.

4 (a) $\mathbf{H}_0 : p = 0.5$

(b) $\mathbf{H}_1 : p \neq 0.5$

(c) The test statistic is X, the number of times a head was tossed in 100 tosses. (Note that you could alternatively have used the number of times a tail was tossed.)

5.2 Finding critical values and critical regions and using significance levels

We start off hypothesis testing by assuming the null hypothesis is true. We then look for evidence to reject or fail to reject it. To do this we need to work out the probability that the test statistic (usually denoted by X) takes particular values.

n is the sample size and X is the number of times the test statistic occurs.

To start off this process we have to decide on the values for n and p so that we can model the situation using a binomial distribution (i.e. $B(n, p)$).

We can then use the distribution to determine probabilities and find critical regions. Critical regions are those regions where if the value of X were to land in them, would cause the null hypothesis $\mathbf{H}_0$ to be rejected.

Suppose we want to check whether a coin is biased in favour of heads or not and conduct an experiment by tossing 10 coins. The null and alternative hypotheses could be $\mathbf{H}_0 : p = 0.5$ and $\mathbf{H}_1 : p > 0.5$

We assume the null hypothesis is true and model the situation using B(10,0.5).

The probabilities of obtaining 0 to 10 heads can be obtained using the formula. These probabilities can be used to produce the following table and graph.

No. of heads	0	1	2	3	4	5	6	7	8	9	10
Probability	0.0010	0.0098	0.0439	0.1172	0.2051	0.2461	0.2051	0.1172	0.0439	0.0098	0.0010

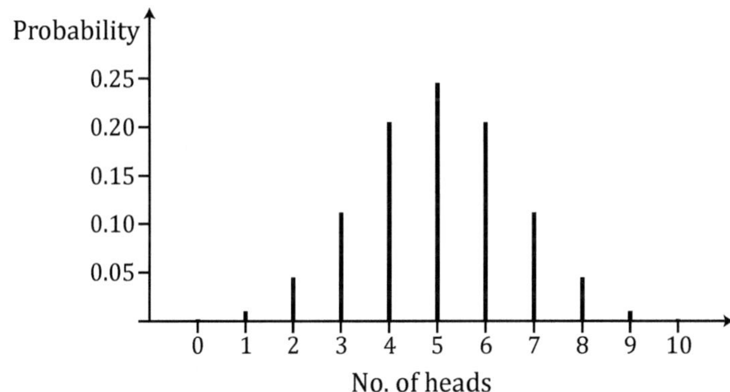

Notice from the graph and the table that the probability distribution is symmetrical and that the most probable number of heads is 5 if the coin is unbiased.

Active Learning

Each of the probabilities in this table have been calculated using the binomial formula:

$$P(X = x) = \binom{n}{x}p^x(1 - p)^{n-x}$$

However, you can calculate all of these values in one go by entering the x-values (i.e. 0 to 10) as a list and then entering the values of n and p (i.e. 10 and 0.5 respectively) using a statistical calculator.

You will then obtain the above list of probabilities. Do this and check your list is the similar to that above.

From the table we can work out the probability of obtaining 'including' and 'greater than' a certain number of heads.

For example, if we want the probability of obtaining 9 or 10 heads we can add the probability of obtaining 9 heads to the probability of 10 heads.

Hence $P(9 \text{ or } 10 \text{ heads}) = P(X \geq 9) = P(9 \text{ heads}) + P(10 \text{ heads})$

$= 0.0098 + 0.0010 = 0.0108$

$P(8, 9 \text{ or } 10 \text{ heads}) = P(X \geq 8) = P(8 \text{ heads}) + P(9 \text{ heads}) + P(10 \text{ heads})$

$= 0.0439 + 0.0098 + 0.0010 = 0.0547$

Note that you could also use a calculator to find these probabilities.

You have already seen that it is possible to use a table of values called the **binomial distribution function table** to find the above probabilities.

The significance level of a test (given the Greek symbol α) tells you how unlikely a value of the test statistic X needs to be before the null hypothesis $\mathbf{H}_0$ is rejected.

The significance level is the probability of the null hypothesis $\mathbf{H}_0$ being rejected even though it is true. It is important to remember that a random sample can contain extreme data which can cause the null hypothesis incorrectly being rejected. In most questions, the significance level will be given and is usually either 1% ($\alpha = 0.01$), 5% ($\alpha = 0.05$) or 10% ($\alpha = 0.1$).

Continuing on from our coin tossing experiment we will use a significance level of 5% which is 0.05 as a decimal. We now need to find the smallest value of X that would make the probability 0.05 or over.

So for $X \geq 9$, $p(X \geq 9) = 0.0108$ which is less than the significance level of 0.05 meaning that if a value of the test statistic X of 9 or 10 was obtained it would mean the null hypothesis should be rejected.

For $X \geq 8$, $p(X \geq 8) = 0.0547$ which is greater than the significance level of 0.05. If the value of the test statistic X is 8 or below (i.e. 8, 7, 6, 5, 4, 3, 2, 1 or 0) the null hypothesis would not be rejected. If a value of X falls into the region outside the critical region the null hypothesis would not be rejected.

The critical value and critical region

The critical region is the set of values of X that would cause the null hypothesis to be rejected. Normally this region is expressed as an inequality such as $X \geq 8$ or $X \leq 2$.

The values of X of 9 or 10 are therefore in the critical region so we can write the critical region as the inequality $X \geq 9$. The values of 0, 1, 2, 3, 4, 5, 6, 7 and 8 lie outside the critical region which means if the test statistic were to lie in this region we would fail to reject the null hypothesis.

Note that in a one-tailed test there is one critical value and with a two-tailed test there are two critical values. This is covered in more depth later.

The critical value is the first value you meet on entering the critical region

In our example the critical value would be 9 as this is the first value you meet when entering the critical region.

The critical value and critical region are shown on the graph at the top of the facing page.

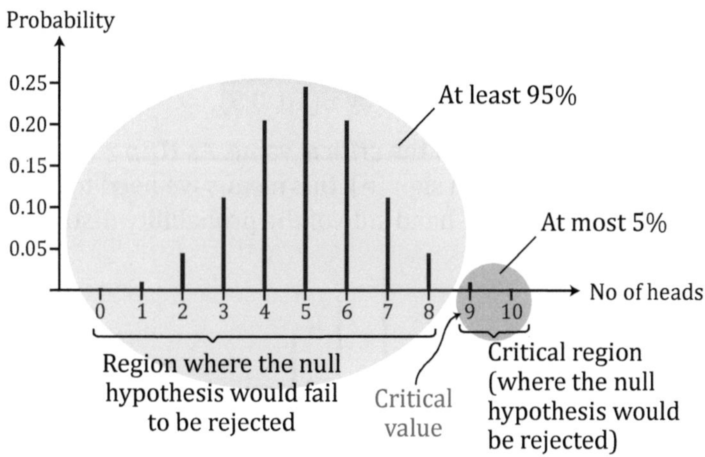

Once the critical value has been found you can compare the value of the test statistic (X) with it to see whether the null hypothesis should be rejected.

This means that if in tossing ten coins and we obtained either 9 or 10 heads at the 5% level of significance we would conclude that the coin was probably biased in favour of heads, which means the null hypothesis is rejected in favour of the alternative hypothesis.

Finding critical regions and values using binomial distribution tables

Calculating the individual probabilities using the binomial formula is tedious especially if there are lots of values of the test statistic X. Luckily there are easier methods using the binomial distribution tables or a calculator.

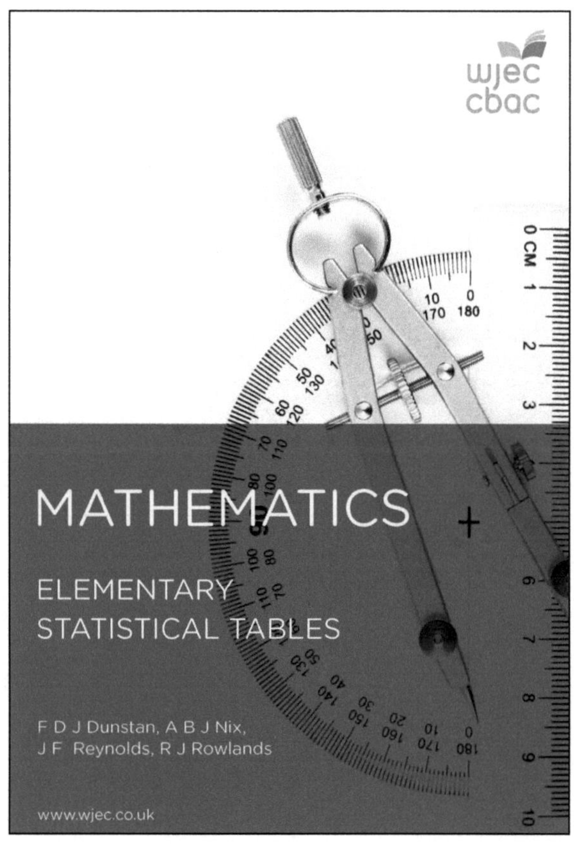

You will be supplied with a copy of a set of elementary statistical tables for your course and you will be given a clean copy in the exam. The table we need to use in the remainder of this topic is 'Table 1 Binomial Distribution Function'.

The above set of tables will be needed for the rest of this topic. Your teacher/ lecturer will give you a copy or you can download a copy from the WJEC website.

The previous example was represented by B(10, 0.5).

Here is how to use the table to find the critical value. As $\mathbf{H}_1: p > 0.5$ the alternative hypothesis contain a greater than sign (>), this means we need to look at the upper tail. The upper tail is at the right-hand side of the probability distribution.

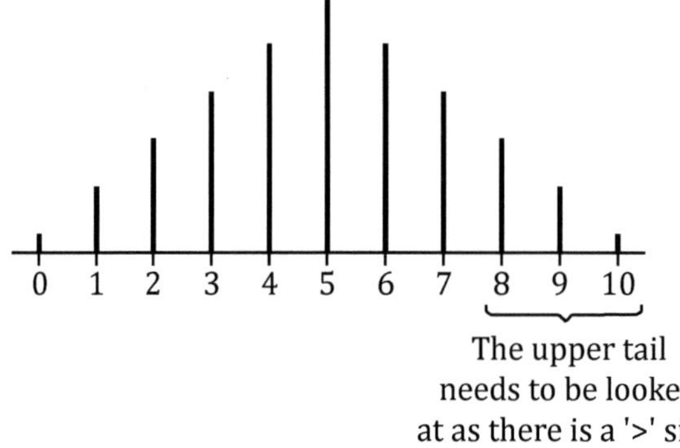
0 1 2 3 4 5 6 7 8 9 10
The upper tail needs to be looked at as there is a '>' sign in the alternative hypothesis.

Now the tables give the probability of at least *x* successes so they give the cumulative probability working from the left tail towards the right tail. This means we deduct the 0.05 from 1 giving a value of 0.95 and then we proceed to use the tables in the following way.

Look at the section of the table with $n = 10$ and then look down the column for $p = 0.5$.

We now need to look down the column and find the first probability in the column that exceeds 0.95 (i.e. 1 – 0.05) and then the critical value is the value of *X* after it.

So here when looking down the column we see the first value exceeding 0.95 is 0.9893 which corresponds to a value of *X* of 8. We need to now pick the value of *X* after this to give the value of *a*. Hence $a = 9$. Hence the critical value is 9 and the critical region is $X \geq 9$.

BINOMIAL DISTRIBUTION FUNCTION

n	x \ p	0.01	0.02	0.03	0.04	0.05	0.06	0.07	0.08	0.09	0.10	0.15	0.20	0.25	0.30	0.35	0.40	0.45	0.50	p / x
n = 9	0	.9135	.8337	.7602	.6925	.6302	.5730	.5204	.4722	.4279	.3874	.2316	.1342	.0751	.0404	.0207	.0101	.0046	.0020	0
	1	.9966	.9869	.9718	.9522	.9288	.9022	.8729	.8417	.8088	.7748	.5995	.4362	.3003	.1960	.1211	.0705	.0385	.0195	1
	2	.9999	.9994	.9980	.9955	.9916	.9862	.9791	.9702	.9595	.9470	.8591	.7382	.6007	.4628	.3373	.2318	.1495	.0898	2
	3	1.000	1.000	.9999	.9997	.9994	.9987	.9977	.9963	.9943	.9917	.9661	.9144	.8343	.7297	.6089	.4826	.3614	.2539	3
	4			1.000	1.000	1.000	.9999	.9998	.9997	.9995	.9991	.9944	.9804	.9511	.9012	.8283	.7334	.6214	.5000	4
	5						1.000	1.000	1.000	1.000	.9999	.9994	.9969	.9900	.9747	.9464	.9006	.8342	.7461	5
	6										1.000	1.000	.9997	.9987	.9957	.9888	.9750	.9502	.9102	6
	7												1.000	.9999	.9996	.9986	.9962	.9909	.9805	7
	8													1.000	1.000	.9999	.9997	.9992	.9980	8
	9															1.000	1.000	1.000	1.000	9
n = 10	0	.9044	.8171	.7374	.6648	.5987	.5386	.4840	.4344	.3894	.3487	.1969	.1074	.0563	.0282	.0135	.0060	.0025	.0010	0
	1	.9957	.9838	.9655	.9418	.9139	.8824	.8483	.8121	.7746	.7361	.5443	.3758	.2440	.1493	.0860	.0464	.0233	.0107	1
	2	.9999	.9991	.9972	.9938	.9885	.9812	.9717	.9599	.9460	.9298	.8202	.6778	.5256	.3828	.2616	.1673	.0996	.0547	2
	3	1.000	1.000	.9999	.9996	.9990	.9980	.9964	.9942	.9912	.9872	.9500	.8791	.7759	.6496	.5138	.3823	.2660	.1719	3
	4			1.000	1.000	.9999	.9998	.9997	.9994	.9990	.9984	.9901	.9672	.9219	.8497	.7515	.6331	.5044	.3770	4
	5					1.000	1.000	1.000	1.000	.9999	.9999	.9986	.9936	.9803	.9527	.9051	.8338	.7384	.6230	5
	6									1.000	1.000	.9999	.9991	.9965	.9894	.9740	.9452	.8980	.8281	6
	7											1.000	.9999	.9996	.9984	.9952	.9877	.9726	.9453	7
	8												1.000	1.000	.9999	.9995	.9983	.9955	.9893	8
	9														1.000	1.000	.9999	.9997	.9990	9
	10																1.000	1.000	1.000	10
n = 11	0	.8953	.8007	.7153	.6382	.5688	.5063	.4501	.3996	.3544	.3138	.1673	.0859	.0422	.0198	.0088	.0036	.0014	.0005	0
	1	.9948	.9805	.9587	.9308	.8981	.8618	.8228	.7819	.7399	.6974	.4922	.3221	.1971	.1130	.0606	.0302	.0139	.0059	1
	2	.9998	.9988	.9963	.9917	.9848	.9752	.9630	.9481	.9305	.9104	.7788	.6174	.4552	.3127	.2001	.1189	.0652	.0327	2
	3	1.000	1.000	.9998	.9993	.9984	.9970	.9947	.9915	.9871	.9815	.9306	.8389	.7133	.5696	.4256	.2963	.1911	.1133	3
	4			1.000	1.000	.9999	.9997	.9995	.9990	.9983	.9972	.9841	.9496	.8854	.7897	.6683	.5328	.3971	.2744	4
	5					1.000	1.000	1.000	.9999	.9998	.9997	.9973	.9883	.9657	.9218	.8513	.7535	.6331	.5000	5
	6									1.000	1.000	1.000	.9997	.9980	.9924	.9784	.9499	.9006	.8262	.7256	6
	7											1.000	.9998	.9988	.9957	.9878	.9707	.9390	.8867	7
	8												1.000	.9999	.9994	.9980	.9941	.9852	.9673	8

We now know the following:

- The critical value is $X = 9$
- The critical region is $X \geq 9$
- The acceptance region is $X \leq 8$

Notice that 9 is the first value reached when going from the acceptance into the critical region. As it is the first value in the critical region it is the critical value.

This means if we obtained 9 heads there is evidence that the null hypothesis should be rejected in favour of the alternative hypothesis.

If we obtained 10 heads there is evidence that the null hypothesis should be rejected in favour of the alternative hypothesis.

Both of the above values lie in the critical region.

If, for example, we obtained 5 heads then this lies in the acceptance region so there is little evidence to reject the null hypothesis. Hence we would fail to reject the null hypothesis.

Finding the critical value and critical region when there is a '<' sign in the alternative hypothesis

When finding the critical value and critical region when there is a '<' sign in the alternative hypothesis you need to investigate the lower-tail of the probability distribution.

The method we use is best described by looking at the following example.

Example

1 The test statistic X has the binomial distribution B(10, 0.40). If the null hypothesis is $\mathbf{H}_0 : p = 0.40$ and the alternative hypothesis is $\mathbf{H}_1 : p < 0.40$. Find the critical region if the significance level is 5%.

Answer

1 Assuming the null hypothesis is true, we use B(10, 0.40).

Using the binomial distribution function table we find the first probability working down the column that exceeds 0.05.

Using the table $P(X \leq 2) = 0.1673$, so a, the critical value, is the value before this (i.e 1). Hence the critical value is 1 and the critical region is $X \leq 1$ so this corresponds to values of X of 0 and 1.

Notice that as the alternative hypothesis is $H_1 : p < 0.40$ we need to look at the lower tail of the distribution.

Note that values of X of 2, 3, 4, 5, 6, 7, 8, 9 and 10 are all in the acceptance region. If a value of the test statistic X from a sample, was found to lie in this region, we would fail to reject the null hypothesis.

Finding the critical value and critical region when there is a '>' sign in the alternative hypothesis

To find the critical value and critical region when there is a '>' sign in the alternative hypothesis, you need to investigate the upper tail of the probability distribution. As the tables give the probability of at least x successes, in order to use the tables it is necessary to subtract the significance level from 1 to find the probability we need to use with the tables. So if the significance level, $\alpha = 0.05$ (i.e. 5%) then we use $1 - \alpha = 1 - 0.05 = 0.95$.

The following example shows how to work out the critical value and critical region for a hypothesis test involving the upper tail.

Examples

1 The test statistic X has the binomial distribution B(10, 0.20), If the null hypothesis is $\mathbf{H}_0 : p = 0.20$ and the alternative hypothesis is $\mathbf{H}_1 : p > 0.20$, find the critical region if the significance level is 5%.

As we are investigating the upper tail we subtract the significance level from 1 to give the probability.

Remember – the value which is on the boundary of the critical region is called the critical value.

We look for the first value of X where the probability exceeds 0.95.

Here $P(X \le 4) = 0.9672$.

Now the critical value is the value before this value of X, so in this case $X = 5$ is the critical value.

Answer

1 As $\mathbf{H}_1 : p > 0.20$ we need to use the upper tail of the probability distribution.

Assuming the null hypothesis is true we use B(10, 0.20) to find values of X with a probability near to 0.95.

$$P(X \le 3) = 0.8791$$

which is less than 0.95, so 4 would lie in the acceptance region.

$$P(X \le 4) = 0.9672$$

which is at least 0.95.

This means $X = 5$ is the critical value and $X \ge 5$ is the critical region. The values in the critical region are 5, 6, 7, 8, 9, 10.

---

2 The test statistic X has the binomial distribution B(30, 0.4), If the null hypothesis is $\mathbf{H}_0 : p = 0.4$ and the alternative hypothesis is $\mathbf{H}_1 : p > 0.4$. Find the critical region and the critical value if the significance level is 1%.

Answer

2 The alternative hypothesis is $\mathbf{H}_1 : p > 0.4$ and as there is a '>' sign in this we need to look at the upper tail of the probability distribution.

Assuming the null hypothesis is true we use B(30, 0.4) and work down the probability column until we find the first value that exceeds 0.99 (i.e. 1 – 0.01).

From the table $P(X \le 18) = 0.9917$ which is greater than 0.99. The critical value, a, is one more than this value so the critical value is 19. Hence $X = 19$ is the critical value and $X \ge 19$ is the critical region.

---

3 Amy is a popular student who would like to be elected as a student school governor. The elections for the student governor are soon and Amy thinks she will have the support of 45% of the students eligible to vote. Amy's best friend thinks that Amy is overestimating her support. Her best friend decides to do a test by taking a random sample of 20 students and asking them who they intended to vote for and 5 said they would vote for Amy.

(a) Write down a suitable test statistic.

(b) Write down suitable null and alternative hypotheses.

(c) Estimate at the 5% significance level, whether Amy is overestimating her support.

Answer

As you are testing in one direction only (i.e. $p < 0.45$), this is a one-tailed test and as there is a '<' sign we will be looking at the lower tail of the probability distribution.

3 (a) The test statistic is X, the number of students who intend to vote for Amy.

(b) The null hypothesis is $\mathbf{H}_0 : p = 0.45$
The alternative hypothesis is $\mathbf{H}_1 : p < 0.45$

(c) Look in the table for $n = 20, p = 0.45$.

Look for the first probability in the column that exceeds 0.05, using the table $P(X \leq 5) = 0.0553$

Now a is the X value before it, so the critical value, $a = 4$

This means the critical region is $X \leq 4$ (note that the critical region inequality is in the same direction as the alternative hypothesis inequality).

Now as the result of the sample is $X = 5$ this lies in the acceptance region meaning that there is not enough evidence to reject the null hypothesis, $\mathbf{H}_0$. So there is evidence to suggest that Amy does have 45% of the vote.

Make sure you make a clear statement about the results in the context of the question.

5.3 One- and two-tailed tests

An alternative hypothesis can be either one tailed or two tailed. Up to now all the hypotheses have been one tailed as they have only looked at one end of the probability distribution (i.e. the upper or the lower tail).

One-tailed hypothesis/one-tailed test

A one-tailed alternative hypothesis is a hypothesis where the parameter being investigated is either less than, or greater than, the value used for the null hypothesis $\mathbf{H}_0$. So, in the example of the coin, an alternative hypothesis might be that the coin is biased in favour of heads, so the one-tailed test would be $\mathbf{H}_1 : p > 0.5$.

If you wanted the alternative hypothesis to be that the coin was biased in favour of tails, then the one-tailed test would be $\mathbf{H}_1 : p < 0.5$ (i.e. the probability of obtaining a head is less because the coin is biased towards tails).

Two-tailed hypothesis/two-tailed test

A two-tailed alternative hypothesis is a hypothesis where the parameter being investigated is not equal to the value used for the null hypothesis $\mathbf{H}_0$ and the test for this is called a two-tailed test.

The two-tailed test would be $\mathbf{H}_1 : p \neq 0.5$ (note that this is made up of the following two tests: $\mathbf{H}_1 : p < 0.5$ and $\mathbf{H}_1 : p > 0.5$).

Finding the critical values and critical ranges for a two-tailed test

For two-tailed tests we divide the significance level by 2 and apply the result to each tail. So if the level of significance was 5% then there would be a 2.5% probability in the lower tail and a 2.5% probability in the upper tail. There would be two critical values and two critical regions in each tail.

The following example shows the technique for finding the critical values and critical regions.

Example

1 The test statistic X has the binomial distribution B(10, 0.50), If the null hypothesis is $\mathbf{H}_0 : p = 0.50$ and the alternative hypothesis is $\mathbf{H}_1 : p \neq 0.50$. Find the critical values and the critical regions if the significance level is 5%.

Answer

1 As this is a two-tailed test we first divide the significance level (i.e. 0.05 in this example) by 2 to give 0.025 at either tail.

First consider the tail to the left and use the binomial distribution function tables to find the first value in the column where the probability exceeds 0.025.

Using the tables, we have $P(X \leq 2) = 0.0547$ as the first value that exceeds 0.025, the critical value is the value before it, so $X = 1$ is the critical value and the critical region for this tail is $X \leq 1$.

Now considering the upper tail we use B(10, 0.50) and the tables to find the first probability in the column that exceeds 0.975 (i.e. 1 – 0.025).

Using the tables, we have $P(X \leq 8) = 0.9893$ so the critical value is the value after it which is $X = 9$. The critical region is $X \geq 9$.

Hence the critical values are 1 and 9 and the critical regions are $X \leq 1$ and $X \geq 9$.

5.4 Interpreting and calculating type I and type II errors

There are two types of error in statistical hypothesis testing which cannot be completely avoided: type I errors and type II errors.

Type I error – A type I error is made when you incorrectly reject a true null hypothesis.
Type II error – A type II error is made when you accept a false null hypothesis.

Here is a summary showing the possible results if the null hypothesis is true or if it is false.

	H_0 is true	**H_0 is false**
Fail to reject H_0	Correct decision	Type II error
Reject H_0	Type I error	Correct decision

Here are some examples of these types of errors to help you understand the differences:

Example 1

A new tablet containing a drug is given to a patient to cure a condition is compared with a tablet containing no drug (called a placebo).

Null hypothesis ($\mathbf{H}_0$) – the patient's condition after taking the drug shows no improvement compared to the placebo.

Alternative hypothesis ($\mathbf{H}_1$) – the patient's condition improves compared to the placebo.

A type I error would indicate that the drug was more effective than the placebo when it actually wasn't. This would involve incorrectly rejecting the null hypothesis.

A type II error would be when the sample indicated that drug showed no patient improvement compared with the placebo, when in fact there was actually an improvement. The null hypothesis would be incorrectly retained.

Example 2

Suppose you want to find if adding fluoride to drinking water reduces tooth decay.

Null hypothesis ($\mathbf{H_0}$) – the addition of fluoride has no effect on tooth decay.

Alternative hypothesis ($\mathbf{H_1}$) – the addition of fluoride reduces tooth decay.

A type II error occurs when the sample fails to detect an effect (i.e. a reduction in the number of fillings when fluoride is added) that is actually present. This means that the experimental data collected dictates the null hypothesis is incorrectly retained/not rejected.

What is the probability of a type I error?

A type I error occurs when we reject a null hypothesis that is true and the probability of a type I error is equal to the level of significance used for the test. It is only possible to determine the probability of a type I error because the null hypothesis is mathematically precise, since the probability distribution is known (i.e. $B(n, p)$) and the parameter involved is known.

With a type II error, the null hypothesis is false and is not rejected. As it is false you don't know if the distribution applies and what parameter is involved.

When you set the significance level, you are setting the maximum probability for a type I error. Unfortunately, as the probability of a type I error decreases, the probability of a type II error increases. Type I and type II errors are part of the process of hypothesis testing and they are impossible to eliminate completely. You could reduce the level of significance from 5% to 1% but although the probability of making a type I error decreases the probability of making a type II error increases.

The only way to reduce the probability of both types of error at the same time, is to use a larger sample which hopefully will be more representative of the population. The problems with an increased sample size are that the samples take longer to collect and cost more.

5.5 Conducting a hypothesis using *p*-values

There is another way of performing a hypothesis test which does not involve finding critical values and critical regions by using probability values, called *p*-values. These *p*-values are obtained from the binomial distribution tables. The values are then compared with the significance levels to determine the significance of the results (i.e. whether to reject or fail to reject the null hypothesis). A *p*-value can take any value between 0 and 1.

We generally interpret *p*-values along the following lines:

$p < 0{\cdot}01$	there is very strong evidence for rejecting $\mathbf{H_0}$
$0{\cdot}01 \le p \le 0{\cdot}05$	there is strong evidence for rejecting $\mathbf{H_0}$.
$p > 0{\cdot}05$	there is insufficient evidence for rejecting $\mathbf{H_0}$.

An Indian takeaway promises customers that at least half of home delivery times are 30 minutes or less. The owner who took A-level maths at school thinks the proportion is lower than this and he decides to test this premise by conducting a hypothesis test.

The test statistic he used is X = number of deliveries taking 30 minutes or less.

The p-value is the probability that the observed result, or a more extreme one, will occur under the null hypothesis $\mathbf{H}_0$.

The null hypothesis is $\mathbf{H}_0 : p = 0.5$ and the alternative hypothesis is $\mathbf{H}_1 : p < 0.5$. The owner samples 30 delivery times and finds that 9 deliveries took 30 minutes or less. He tests at the 5% level of significance and here are his findings:

$$p(X \leq 9) = 0.0214$$

The p-value is 0.0214 which is less than the significance level of 0.05. The p-value means that there is a probability of 0.0214 the owner will mistakenly reject the premise that 50% of home delivery times are 30 minutes or less.

We now look at the p-value table and we compare the p-value to see which range it fits into $0{\cdot}01 \leq p \leq 0{\cdot}05$, so there is strong evidence for rejecting $\mathbf{H}_0$.

The owner concludes that there is strong evidence that the premise is incorrect, but there is always the possibility that the sample collected was not representative of the population.

α, the significance level is usually set at either 5% (i.e. 0.05) or 1% (i.e. 0.01).

Here are the steps needed to conduct a significance test using p-values:

1 Write the null and alternative hypotheses.

2 Specify the test statistic you are going to use.

3 Assume that the null hypothesis is true and decide on the value of the test statistic (usually this is given in the question).

4 Use the distribution of the test statistic (in this topic it will be $B(n, p)$) and obtain the probability using tables. The probability is the probability of observing a more extreme test statistic in the direction of the alternative hypothesis than we did.

5 Compare the probability to the significance level α. If the probability is less than or equal to α, the null hypothesis is rejected in favour of the alternative hypothesis. If the probability value is greater than α, the null hypothesis is not rejected.

Examples

1 By past experience a company has a $\frac{3}{10}$ chance of winning a contract. The managing director has had previous experience in using new computer software and experiments conducted indicated that with using the new computer software they can win 9 out of 20 contracts. He says this will be an improvement.

Using a 5% significance level, test the claim made by the managing director.

Answer

1 In the following answer we have used the steps as outlined above. When answering questions you do not have to write each step out – just the 'working out' below the steps.

Step 1: Write the null hypothesis and alternative hypothesis.

p is the probability of winning an individual contract.

$\mathbf{H}_0 : p = 0.3$

$\mathbf{H}_1 : p > 0.3$

Notice the word 'chance' in the question. This means that we know the probability of winning a contract p is 0.3.

Step 2: Specify the test statistic to be used.

X is the number of contracts won using the new computer software.

It is important to specify the model for the test statistic. In this topic we will only be using the binomial distribution (i.e. $B(n, p)$).

Step 3: Assuming $\mathbf{H}_0$ is true, calculate the probability of winning 9 or more contracts. Note that as there is a '>' sign in the alternative hypothesis you are looking at the upper tail of the probability distribution.

Step 4: Using $B(n, p)$, calculate the probability of observing a more extreme value of the test statistic in the direction of the alternative hypothesis (i.e. in the upper tail in this case).

X is $B(20, 0.3)$

$P(X \geq 9) = 1 - P(X \leq 8)$

$= 1 - 0.8867$

$= 0.1133$

The binomial tables cover values of at most x successes. So if we want greater than or equal to 9 successes we need to subtract the probability of 8 or fewer successes, from 1.

Step 5: Compare the probability value to the significance level α

$\alpha = 5\% = 0.05$

$0.1133 > 0.05$ so as the p-value is greater than α, there is insufficient evidence to reject the null hypothesis.

0.1133 is the p-value. Don't get the p-value mixed up with the binomial probability of success for each trial p.

This means that there is evidence that the new software does not improve winning contracts, so it is not an improvement.

Here you must make a statement based on your hypothesis test. This has to be in the context of the question.

---

2 An older drug to treat diabetes has a probability of success of $\frac{3}{10}$ in patients with the disease. Research has been undertaken and a new drug has been developed which has been successful with 19 out of 50 patients with the disease. The drug company who make the new drug say that taking the new drug offers an improvement compared to the old drug.

(a) Write down suitable null and alternative hypotheses.

(b) Describe a suitable test statistic that should be used.

(c) Write down the distribution of the random variable X that represents the number of patients successfully treated by the drug if the null hypothesis was true.

(d) Using tables or otherwise, find the probability of X taking a value equal to or greater than 19.

(e) A significance test is carried out with a significance level of 5%. Using this significance level does 19 out of 50 supply evidence that the null hypothesis $\mathbf{H}_0$ should be rejected? Explain your answer.

Answer

2 For this answer we will just find probabilities using the tables and not determine the critical value and critical region. As no method is indicated in the question, this provides a different way of answering this question.

$\mathbf{H}_0$: The new drug offers the same probability of success as the old drug.

$\mathbf{H}_1$: There is a greater probability of success with the new drug.

The test statistic uses the number of successes in the sample.

Remember that the tables give a probability of less than, or equal to a certain value of X.

(a) $\mathbf{H}_0 : p = \frac{3}{10}$.

$\mathbf{H}_1 : p > \frac{3}{10}$

(b) The test statistic is X, the number of patients whose condition was improved by taking the new drug.

(c) X is B(50,0.3)

(d) $P(X \geq 19) = 1 - P(X \leq 18)$

$= 1 - 0.8594$

$= 0.1406$

(e) As $0.1406 > 0.05$, there is not enough evidence to reject the null hypothesis $\mathbf{H}_0$.

This means that there is evidence to suggest that the new drug is no better than the old drug.

BOOST Grade

It is important to remember that a hypothesis cannot provide concrete proof – it can only supply evidence.

Active Learning

Instead of answering parts (d) and (e) of the above question in the way shown, redo the answer, but this time find the critical values and critical region.
Evaluate which method you prefer.

Finding the significance level from the critical value

If you know the critical value, then you can work backwards from it and work out what the significance level is. Remember that the significance level is the probability of making a type I error (i.e. the error made when you incorrectly reject a true null hypothesis). In order to determine the probability of a type II error you would need to know the exact probability and the following examples shows the method used.

Examples

1 The test statistic X has the binomial distribution B(20, 0.30). If the null hypothesis is $\mathbf{H}_0 : p = 0.30$ and the alternative hypothesis is $\mathbf{H}_1 : p < 0.30$.

If the critical region is defined as $X \leq 2$:

(a) Determine the significance level of this critical region.

(b) Explain what is meant by a type II error.

(c) It is found that the actual probability is not 0.30 as thought, but is now 0.25. Calculate the probability of a type II error.

Answer

1 Under the null hypothesis, $\mathbf{H}_0$, X is B(20, 0.30)

(a) Significance level, α, = $P(X \leq 2) = 0.0355$

(b) A type II error is the error that occurs when you retain (i.e. fail to reject) a null hypothesis that is actually false.

(c) X is B(20, 0.25)

Type II error probability = $P(X \geq 3) = 1 - P(X \leq 2) = 1 - 0.0912 = 0.9088$

Here the significance level is the probability of obtaining a value for the test statistic X that is less than or equal to the critical value. Note that it is ≤ because of the < in the alternative hypothesis.

Notice the value of p has changed to 0.25.

2 Dewi, a candidate in an election, believes that 45% of the electorate intend to vote for him. His agent, however, believes that the support for him is less than this. Given that p denotes the proportion of the electorate intending to vote for Dewi,

(a) State hypotheses to be used to resolve this difference of opinion. [1]

They decide to question a random sample of 60 electors. They define the critical region to be $X \leq 20$, where X denotes the number in the sample intending to vote for Dewi.

(b) (i) Determine the significance level of this critical region.

(ii) If in fact p is actually 0.35, calculate the probability of a type II error.

(iii) Explain in context the meaning of a type II error.

(iv) Explain briefly why this test is unsatisfactory. How could it be improved while keeping approximately the same significance level? [8]

Answer

2 (a) $\mathbf{H}_0 : p = 0.45$

$\mathbf{H}_1 : p < 0.45$

This means we need to look at the lower tail of the probability distribution.

(b) (i) Under the null hypothesis, $\mathbf{H}_0$, X is B(60, 0.45).

If you look at the binomial distribution tables you will notice that the values for n only go up to 50. In this question $n = 60$ so you need to use a calculator to find the probabilities.

Significance level = $P(X \leq 20) = 0.0446$

Note that here we need to find the probability that corresponds to $X \leq 20$ as the level of significance is the probability of X being in the critical region.

(ii) X is B(60, 0.35)

Type II error probability = $P(X \geq 21)$

$= 1 - P(X \leq 20)$

$= 1 - 0.4516$

$= 0.5484$

(iii) A type II error here is accepting that support for Dewi is 45% (i.e. the null hypothesis is not rejected) when it is actually 35% so it is false and should be rejected.

BOOST

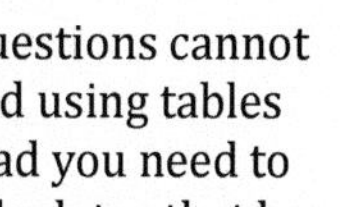

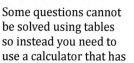

Some questions cannot be solved using tables so instead you need to use a calculator that has a statistical distribution function.

A type II error is a made when a null hypothesis is retained that is actually false. If $X \geq 21$ the null hypothesis would be retained so the type II error probability is $P(X \geq 21)$.

Remember: A type II error is failing to reject a null hypothesis that is false. You must however explain it in the context of the question.

(iv) It is a large value for an error probability. It could be reduced by taking a larger sample (60 people is a very small sample for an election).

3 Say whether each of the following statements concerning hypothesis testing is true or false.

(a) A hypothesis is a statement we make about a statistical population which when supplied with evidence we either fail to reject or reject.

(b) The null hypothesis is always called $\mathbf{H}_1$.

(c) The alternative hypothesis always contradicts the null hypothesis.

(d) A test on a sample of the population is used to either reject, or fail to reject, the null hypothesis.

(e) The critical value is the value where you accept the null hypothesis.

(f) The critical region contains those values that would cause the null hypothesis to be rejected.

(g) The higher the significance level, the stricter the hypothesis test.

(h) The significance level is the probability at which you decide to be satisfied that an event has not occurred by chance

Answer

3 (a) True

(b) False

(c) True

(d) True

(e) False

(f) True

(g) False

(h) True

Test yourself

1. It is suspected that a coin is biased towards heads.
   (a) Write down a suitable test statistic.
   (b) Write down a suitable:
      (i) Null hypothesis
      (ii) Alternative hypothesis.
   (c) If the coin is tossed 9 times, how many heads would there need to be at a level of significance of 5% for the null hypothesis to be rejected?

2. Asha suspects that a five-sided spinner with numbers 1 to 5 is biased towards the number 5. She decides to conduct an experiment to see if this is the case. The spinner was spun 10 times and the number 5 came up 4 times. Conduct a significance test at the 5% level of significance to determine whether the spinner is biased towards the number 5.

3. The probability of a call being answered at a call centre in 5 minutes or more is 0.4 . The manager says that there has been a decrease in the number of callers who have to wait more than 5 minutes. He conducts an experiment and records that for the next 20 calls, 4 calls waited more than 5 minutes. Is there support at the 5% level of significance that the manager is correct?

4. Jack plays a computer game and the probability of him winning is 0.4. He reckons that with practice he has improved his probability of winning. In the next 8 games he plays he wins 6 of them. Is there support at the 5% level of significance that he has improved?

5. A head teacher of a school believes that at least 50% of students get at least one hour of homework per night on a school day. The deputy head teacher thinks this figure is lower than 50%.
   The head teacher decides to test his belief by asking a group of A-level physics students in a lesson he is taking. He asks all 10 students in the class to tell him if they had at least one hour's homework on each day in the previous week.
   (a) (i) Give the name of the sampling method used by the head teacher.
      (ii) Describe a weakness of this sampling method.
   (b) (i) The head teacher decides to test his hypothesis using the results from his sample.
      Write down a null hypothesis and an alternative hypothesis he could use.
      (ii) Find the probability that in the class of ten students, two or fewer students get at least one hour's homework per night.
      (iii) Find the probability that in the class of ten students fewer than four get at least one hour's homework per night.
   (c) Assuming that the 10 physics students are a random sample and that 2 students got more than one hour's homework per night in the previous week, find using a hypothesis test at the 5% level of significance if the head teacher's belief is supported by the evidence.

6 Alex is a candidate in the forthcoming local elections. At the moment Alex claims she has 35% of the vote but her partner thinks she might have less than this. Her partner conducts a short survey. He asks 20 people in the constituency who they would vote for in the election and out of 20 people asked, 6 say they intend to vote for Alex.

(a) Write down suitable null and alternative hypotheses that could enable her to test her claim.

(b) State the distribution of the random variable X that represents the number who would vote for her if the null hypothesis were true.

(c) If the significance level is 5%, find:

 (i) the critical value

 (ii) the critical region.

(d) Using your answer to part (c), does the fact that 6 out of 20 people said they would vote for her in her sample mean the null hypothesis would need to be rejected? Give a reason for your answer.

7 A footballer thinks he has a 45% chance of scoring a goal from a penalty. His manager thinks he is better than this. He looks up his goal scoring records and notes that in 15 penalties he scored 10 goals. Test, at the 5% significance level, whether the player has underestimated his chance of scoring.

8 In a batch of LED light bulbs the production manager claims that there is a 15% chance of one of them being faulty during manufacture. The quality control manager says that this overestimates the probability of a faulty bulb. A test is performed using a random sample of 50 bulbs and 3 were found to be faulty. At the 5% level of significance, test to see if the production manger is correct in her claim that 15% of bulbs are faulty.

(a) Write down suitable null and alternative hypotheses that could enable her to test her claim.

(b) At the 5% level of significance, test to see if the production manger is correct in her claim that 15% of bulbs are faulty.

9 Hope is a car salesperson who tries to sell new cars to customers who have booked test drives.
Hope believes that she sells new cars to 43% of customers who book test drives. Her manager notes from records that out of the last 30 test drives she sold 18 new cars and he believes that the percentage is actually higher than she thinks.
Test Hope's claim at a 1% level of significance.

10 A company manufactures touch screens. It uses state of the art technology, but the quality control manager claims that 25% of the screens produced are faulty.
In order to test this claim, the number of faulty screens in a random batch of 50 screens is recorded.

(a) Give **two** reasons why this situation can be modelled using a binomial distribution.

(b) If the null hypothesis is that 'the probability of a screen being faulty is 0.25', find at the 5% level of significance the critical region for a two-tailed test.

Summary

Check you know the following facts:

Hypothesis testing

Hypothesis testing is used to test a hypothesis about the probability of the number of times (X) a certain property crops up.

The test statistic X is modelled by a binomial distribution $B(n, p)$ where p is the probability of the event occurring in one trial and n is the total number of trials.

Null and alternative hypothesis

Null hypothesis $\mathbf{H}_0 : p =$ a value

Alternative hypothesis for a one-tailed test is $\mathbf{H}_1 : p <$ a value or $\mathbf{H}_1 : p >$ a value

Alternative hypothesis for a two-tailed test is $\mathbf{H}_1 : p \neq$ a value

Significance level

Assuming the null hypothesis is true, the significance level indicates how unlikely a value needs to be before the null hypothesis $\mathbf{H}_0$ is rejected.

The significance level for this topic can be:

1% ($\alpha = 0.01$), 5% ($\alpha = 0.05$) or 10% ($\alpha = 0.1$).

The critical value and critical region

Here you use the significance level along with the values of n and p to find the set of values of X that would cause the null hypothesis to be rejected.

In all of the following cases the significance level is taken as 5% but this can be changed to 1% or 10% or any other significance level.

- If $\mathbf{H}_0$ contains a < sign the critical region will be the lower 5% of values of X (i.e. in the lower tail) so use the tables and find the first p-value in the column which exceeds 0.05 and choose the value of X **before** it. This is the critical value, a, and the critical region will be $X \leq a$.

- If $\mathbf{H}_0$ contains a > sign the critical region will be the upper 5% of values of X (i.e. in the upper tail) so use the tables and find the first p-value in the column which exceeds 0.05 and choose the value of X **after** it. This is the critical value, a, and the critical region will be $X \geq a$.

- If $\mathbf{H}_0$ contains a ≠ sign, the critical region will be the upper 2.5% and lower 2.5% of values of X (i.e. in the upper tail and lower tails). We use the techniques outlined in the previous two paragraphs to find the two critical values and the critical regions in each of the two tails.

Type I and type II errors

A **type I error** is made when you incorrectly reject a true null hypothesis.

A **type II error** is made when you fail to reject a false null hypothesis.

p-values

The p-value is the probability that the observed result or a more extreme one will occur under the null hypothesis.

Assuming that the probability distribution of X is $B(n, p)$ the probability of obtaining a value (called the p-value) of the test statistic or a more extreme one can be found using tables. If this value is less than or equal to the level of significance, the null hypothesis is rejected.

6 Quantities and units in mechanics

Introduction

This is a very short topic that looks at quantities and units in mechanics which will be used in all of the mechanics topics for both AS and A2.

The international system of units (called SI units for short) uses seven fundamental quantities with their units. You only need to understand and use three of them for mechanics and they are length, time and mass. There are other quantities with their own units that can be derived from the fundamental quantities.

This topic covers the following:

6.1 Fundamental quantities and units in the SI system

There are seven fundamental quantities in the SI system of units, but for AS Maths, we only need to know about the three shown in the table:

Fundamental quantity	Unit	Symbol
Length	metre	m
Mass	kilogram	kg
Time	second	s

6.2 Using derived quantities and units

The SI fundamental quantities are the building blocks of the SI system and all other quantities are derived from them. The derived quantities are formed by powers, products or quotients of the fundamental quantities.

For example, velocity is derived from the fundamental quantities length (i.e. distance) divided by time, so the formula is:

$$\text{velocity} = \frac{\text{length}}{\text{time}} \text{ and the units will be } \frac{\text{m}}{\text{s}} = \text{m s}^{-1}$$

Notice the way the rules of indices are used here to obtain the final unit for acceleration.

Sometimes a completely new unit is created. For example, acceleration is the change in velocity divided by the time and has the units of ms^{-2}:

$$\text{acceleration} = \frac{\text{change in velocity}}{\text{time}} \text{ and the units will be } \frac{\text{m s}^{-1}}{\text{s}} = \text{m s}^{-2}$$

Force is a derived quantity as it is derived from the fundamental quantities mass, length and time.

Force is derived using the formula:

$$\text{force} = \text{mass} \times \text{acceleration}$$

and in terms of the fundamental quantities it would have the units kg m s^{-2}. However, force has its own unit of newtons (N). Hence, we can say $1\text{ N} = 1\text{ kg m s}^{-2}$.

Weight is the force of gravity acting on mass and can be derived using the formula:

$$\text{weight} = \text{mass} \times \text{acceleration due to gravity}$$

Weight has the unit N or in terms of the fundamental quantities kg m s^{-2}

There are many derived quantities and the ones you need to know about are shown in the following table:

Derived quantity	Name	Symbol
Area	metres squared	m^2
Volume	metres cubed	m^3
Density	kilograms per metre cubed	kg m^{-3}
Velocity	metres per second	m s^{-1}
Acceleration	metres per second squared	m s^{-2}
Force	newton	N
Weight (Force of gravity)	newton	N
Moment	newton metres	N m

Changing units

It is sometimes necessary to change units so the units are expressed in terms of the fundamental units.

For example, speed/velocity is often expressed in km h^{-1} and often we need to express it in the units m s^{-1}.

To convert 5 km h^{-1} we take the following steps:

- First change the distance to m, so 5 km = 5 × 1000 m = 5000 m
- Then change the time from hours to seconds. There are 60 × 60 = 3600 seconds in one hour.

Hence $5 \text{ km h}^{-1} = \frac{5000}{3600} \text{ m s}^{-1} = 1.39 \text{ m s}^{-1}$ (2 d.p.)

Test yourself

There will be no exam questions purely on this topic as knowledge of this material will be incorporated into all the unit topics in mechanics.

Summary

Check you know the following facts:

The fundamental quantities and their units are:

Length (m)

Time (s)

Mass (kg)

Derived quantities are quantities that are derived from the fundamental quantities using a formula and include the following:

$$\text{velocity} = \frac{\text{length or distance}}{\text{time}}\ (\text{m s}^{-1})$$

$$\text{acceleration} = \frac{\text{change in velocity}}{\text{time taken}}\ (\text{m s}^{-2})$$

$$\text{force} = \text{mass} \times \text{acceleration (N)}$$

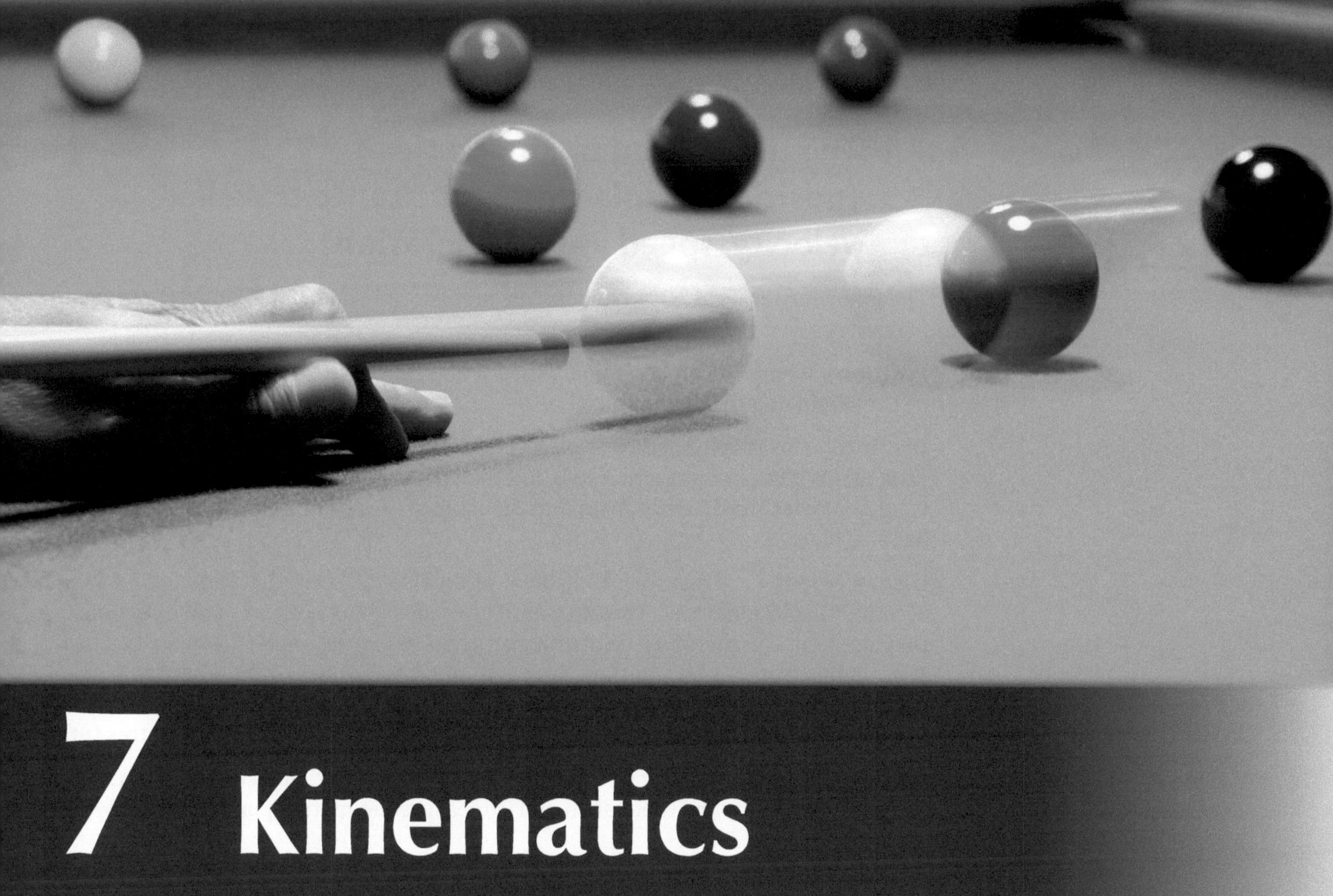

7 Kinematics

Introduction

Kinematics deals with the motion of objects without considering the masses or forces that produce the motion. In this topic you will be considering motion in a straight line under either constant acceleration or variable acceleration. This topic will allow you to understand and produce graphs representing motion in a straight line and use the graphs to find values of velocities, accelerations and distances. You will also be using the equations of motion that you may have used before in GCSE Physics. You will also deal with problems involving acceleration that is not constant and in these problems calculus (differentiation and integration) has to be used.

This topic covers the following:

7.1 The terms used in kinematics (position, displacement, distance travelled, velocity, speed and acceleration) and modelling assumptions

7.2 Displacement–time graphs

7.3 Velocity–time graphs

7.4 Sketching and interpretation of velocity–time graphs

7.5 Using the equations of motion (i.e. the *suvat* equations)

7.6 Applying the *suvat* equations to vertical motion under gravity

7.7 Using calculus in kinematics

7.1 The terms used in kinematics (position, displacement, distance travelled, velocity, speed and acceleration)

The following terms are used to describe motion and you need to understand their meanings:

Position – where the object is positioned with reference to a fixed point which can be the origin or a fixed point on a line.

Distance – this is the length that is travelled and is measured in metres (m). If you walked in a straight line from A to B and then back to A, then the distance travelled would be twice the distance from A to B. Distance is a scalar quantity because it has size only.

Displacement – is a measure of distance but it also takes the direction into account. Displacement is a vector quantity because it has both size and direction. Displacement can therefore be positive or negative. For example, if you walked in a straight line from A to B and then back to A, the displacement would be zero. This is because the displacement one way would be positive and the displacement in the opposite direction would be negative. Displacement is measured in metres (m).

Speed can also be measured in km h^{-1}. To change these speeds into m s^{-1}, multiply by 1000 (i.e. the number of metres in one kilometre) and divide by 3600 (i.e. the number of seconds in one hour). Hence:

$$40 \text{ km h}^{-1} = \frac{40 \times 1000}{3600}$$
$$= 11.1 \text{ m s}^{-1} \text{ (3 s.f.)}$$

Speed – is a scalar quantity so it has size but no direction. Speed is the distance travelled divided by the time taken and is measured in metres per second (m s^{-1}).

Hence $$\text{speed} = \frac{\text{distance travelled}}{\text{time taken}} \quad \text{or using symbols,} \quad s = \frac{d}{t}$$

Velocity – is a vector quantity so it has both size and direction. If the velocity is uniform, then velocity is the change in displacement divided by the change in time (i.e., the rate of change of displacement with time) and is measured in metres per second (m s^{-1}).

A velocity in one direction can be taken as positive so the velocity in the opposite direction would be negative.

Hence $$\text{velocity} = \frac{\text{change in displacement}}{\text{change in time}}$$

Acceleration – is a vector quantity and is the change in velocity divided by the change in time. It is measured in metres per second squared (m s^{-2}). A negative acceleration is called a deceleration or retardation.

Hence $$\text{acceleration} = \frac{\text{change in velocity}}{\text{change in time}}$$

Active Learning

Use the words displacement, velocity and acceleration to describe the motion of a particle in a couple of sentences.

Modelling assumptions

A model is a mathematical representation of a real situation. In this topic you will come across a variety of situations that can be modelled using graphs, equations or both.

When using mathematical models, we need to keep things simple so certain assumptions are made:

- A body (e.g. car, plane, boat, train, stone, ball, etc.) can be modelled as a particle. A particle has dimensions which are negligible but it still has a mass.

- Air resistance and other frictional forces are negligible and therefore are taken to be zero.
- Gravity acts straight down, at right angles to the Earth's surface, which is assumed to be flat.
- Gravity does not depend on height (this is a realistic assumption over small heights where the actual change is small).

7.2 Displacement–time graphs

Displacement–time graphs depend on the type of motion.

Object is stationary

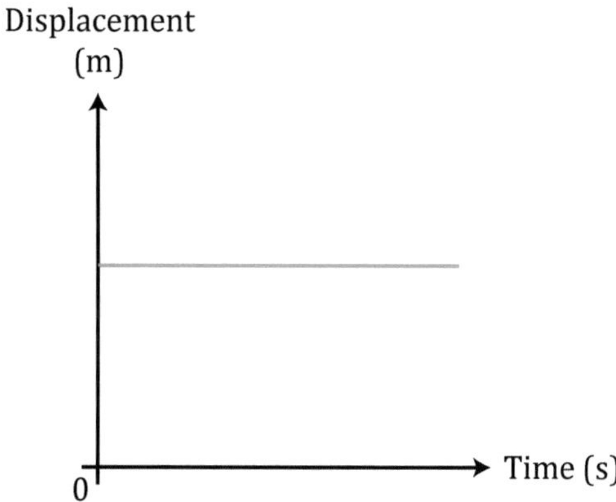

The gradient of a displacement–time graph represents the velocity.

Notice the displacement stays the same with time, so the object is stationary. The gradient of the line is zero and this represents zero velocity.

Object moving with constant velocity

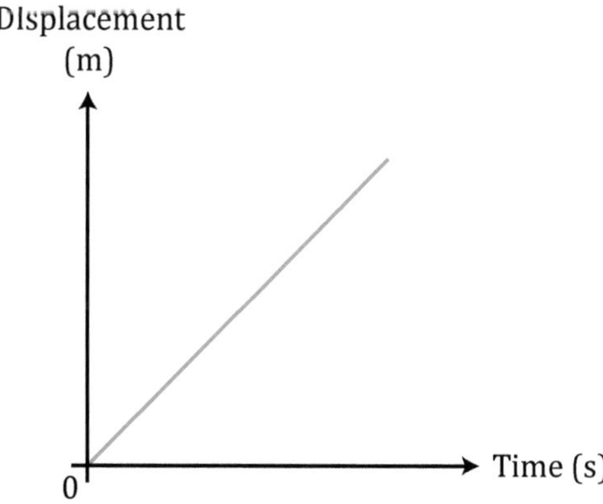

The gradient is positive, so the velocity is positive.

The gradient of the line is constant and represents constant velocity. Constant velocity means the acceleration is zero.

Object accelerating

The gradient of this curve changes with time so the velocity changes.

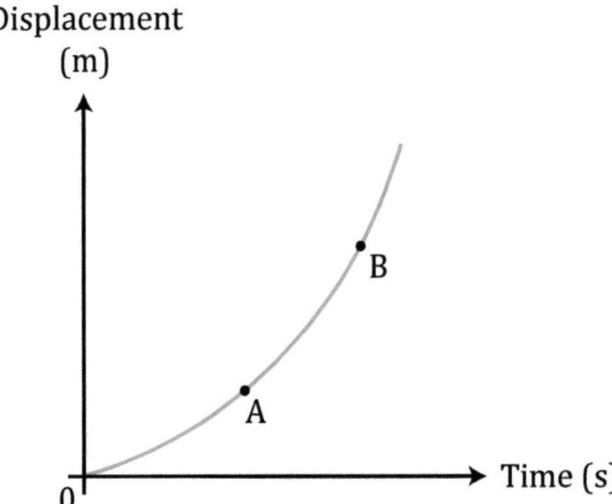

The gradient of the curve at A is smaller than the gradient at B. As the gradient represents the velocity, the velocity is increasing, showing the object is accelerating.

Object decelerating

The gradient and hence the velocity decreases with time.

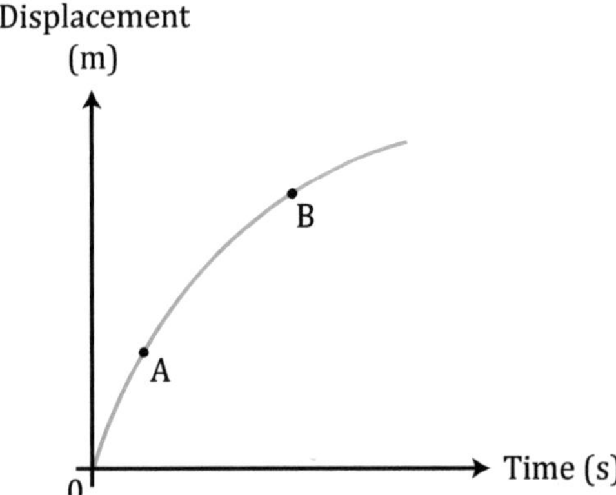

The gradient at A is greater than that at B showing that the velocity of the object is decreasing. The object is therefore decelerating.

7.3 Velocity–time graphs

Velocity–time graphs depend on the type of motion.

Object moving with constant velocity (i.e. no acceleration)

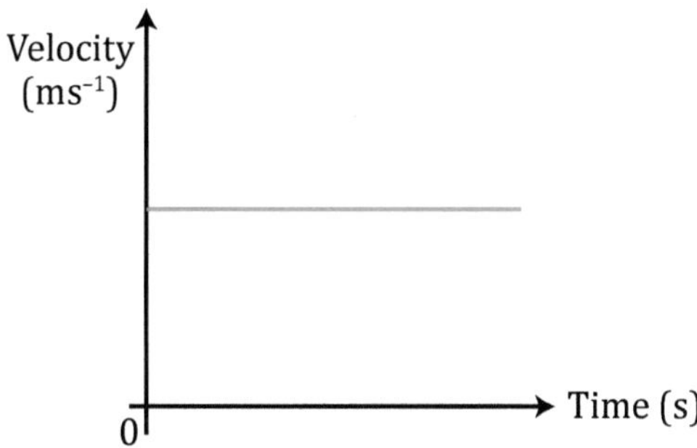

The graph is a straight horizontal line showing that the velocity remains constant with time. The gradient of the line is zero so the acceleration is zero. Horizontal lines represent constant velocity.

Object moving with constant acceleration

A straight line with a positive gradient represents motion with constant acceleration. In the velocity–time graph shown below, the object starts with a velocity u and then accelerates with a constant acceleration to a higher velocity v.

The gradient represents the acceleration. From the graph the acceleration $a = \frac{v - u}{t}$ where u is the initial velocity, v is the final velocity and t is the time.

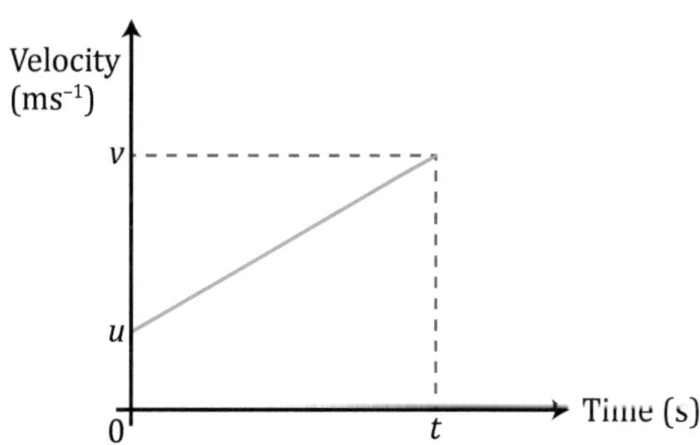

Objects moving with constant deceleration

This velocity–time graph represents motion with constant deceleration as it is a straight line with a negative gradient. Hence the object starts with an initial velocity u and then decelerates to a lower velocity v.

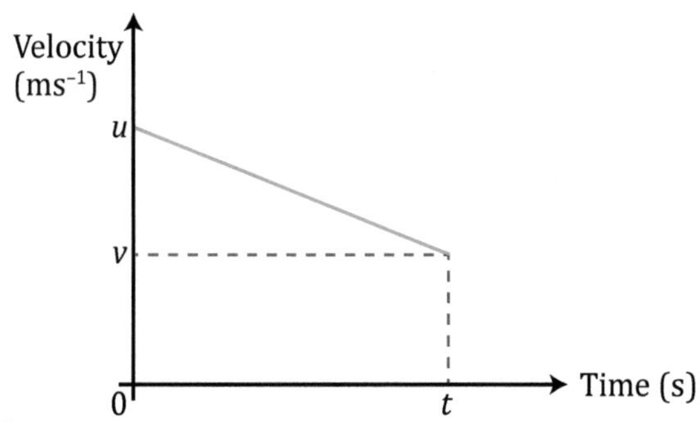

7.4 Sketching and interpretation of velocity–time graphs

The following graph represents a journey:

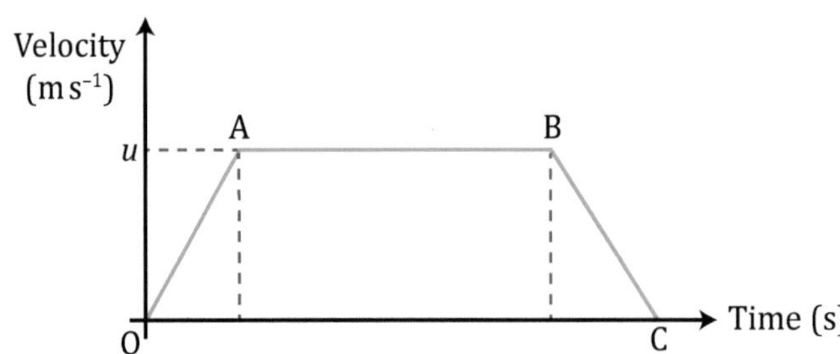

Area of trapezium = $\frac{1}{2}$(sum of the two parallel sides) × perpendicular distance between them.

The gradient of line OA represents constant acceleration. The gradient of line BC is negative showing constant deceleration.

Line AB represents travelling at constant velocity.

The area under the line represents the distance travelled or displacement. From the graph, distance travelled/displacement = area under the graph (i.e. the area of the trapezium OABC). If the area is under the time axis, the area represents a negative displacement.

Examples

1 A particle starting from rest and travelling in a straight line, accelerates uniformly for 2 s and reaches a constant velocity of u m s^{-1}. It travels at u m s^{-1} for 10 s before decelerating uniformly to rest in 3 s. The total distance travelled by the particle was 50 m.

(a) Draw a velocity–time graph to show the motion of the particle.

(b) Find the value of u.

(c) Find the magnitude of the deceleration.

When drawing a velocity–time graph, ensure the axes are labelled with quantities and units. Mark any values and letters for quantities which need to be found, on the graph.

Answer

1 (a)

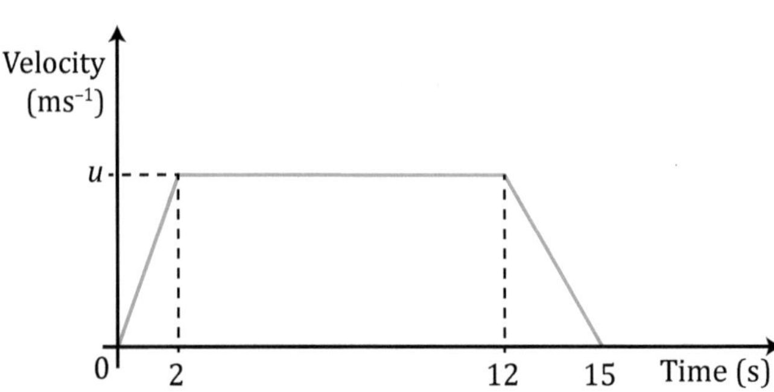

The formula for the area of a trapezium is used here.

(b) Total distance travelled = area under the velocity–time graph

$$= \frac{1}{2}(15 + 10) \times u$$

But the total distance travelled = 50 m

Hence $50 = \frac{1}{2}(15 + 10) \times u$ so $u = 4$ m s^{-1}

This is the gradient of the line between $t = 12$ and $t = 15$ s. Note that a minus sign is not included as we have said it is a deceleration.

(c) Deceleration = $\frac{4}{3} = 1.33$ m s^{-2}

2 The velocity–time graph shown below represents the four stages of motion of a vehicle moving along a straight horizontal road. The initial velocity of the vehicle is 20 m s^{-1}.

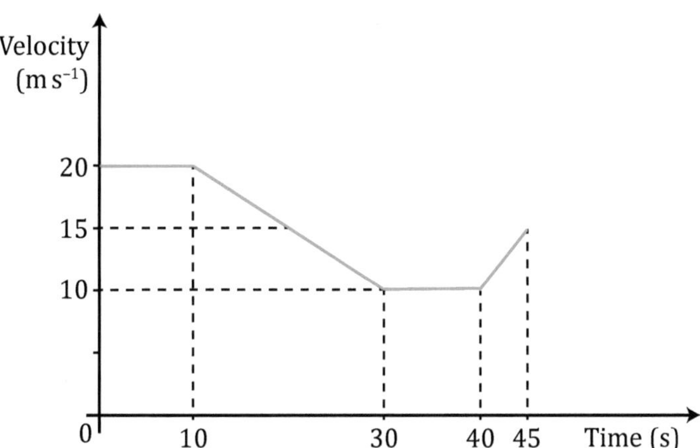

(a) Find the distance travelled whilst the car is decelerating.

(b) Find the total distance travelled whilst travelling at constant speed.

(c) During the later stage of the motion the car accelerates. Calculate the magnitude of the acceleration.

(d) Calculate the total distance travelled during the motion described by the graph.

Answer

2 (a) Distance = area of trapezium

$= \frac{1}{2}(20 + 10) \times 20$

$= 300$ m

(b) Total distance whilst travelling at constant speed = 20 × 10 + 10 × 10

= 300 m

(c) Acceleration = gradient = $\frac{15 - 10}{5} = 1$ m s^{-2}

(d) Distance travelled whilst accelerating = $\frac{1}{2}(10 + 15) \times 5 = 62.5$ m

Total distance travelled = 300 + 300 + 62.5 = 662.5 m

Area of trapezium = $\frac{1}{2}$(sum of the two parallel sides) × distance between them.
You could alternatively, divide the shape into a triangle and a rectangle and add the two areas together. Note you will not be given the formula for the area of a trapezium, so it will need to be remembered.

Note: as you are asked for the *magnitude* of the acceleration you do not need to give its direction.

3 Cars A and B are travelling a long straight road. At time $t = 0$, Car A is travelling with a speed of 20 m s^{-1} and at this time it overtakes Car B travelling with a speed of 15 m s^{-1}. Car B immediately accelerates uniformly and both cars travel a distance of 600 m before Cars A and B are level and pass again.

(a) Draw a velocity–time graph showing the motion of the cars from where they are first level to when they are level again.

(b) Show that the time between overtaking the first and second time is 30 s.

(c) Calculate the magnitude of the velocity of Car B after 30 s.

(d) Calculate the acceleration of Car B.

Answer

Note that point P relates to Car B's velocity being equal to Car A's velocity **not** where Car B overtakes Car A.

3 (a)

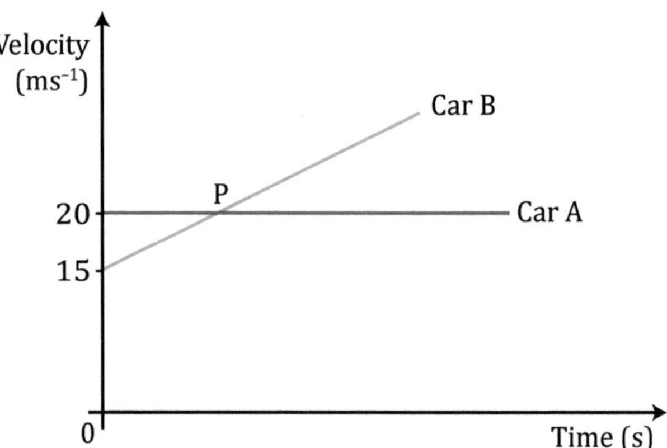

(b) Considering the motion of Car A, suppose the cars are level after time t seconds.

The area under the velocity–time graph for Car A = 600

This is the area under the graph, which is a rectangle of length 20 and width t.

Hence, $20 \times t = 600$

giving $t = 30$ s

(c) After 30 s the distance travelled by Car B is 600 m

Let the velocity of Car B after 30 s = v

Area under the graph = area of a trapezium = $\frac{1}{2}(15 + v) \times 30$

Now, distance travelled = area under the graph

Hence $600 = \frac{1}{2}(15 + v) \times 30$

Solving this equation gives $v = 25\ \text{m s}^{-1}$

(d)

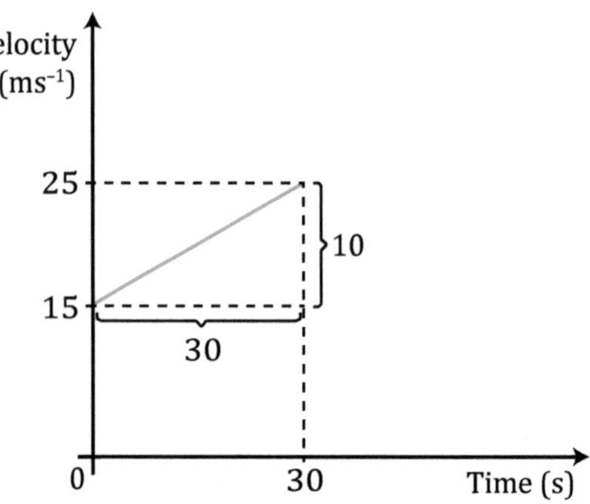

Acceleration = gradient = $\frac{10}{30} = 0.33\ \text{m s}^{-2}$

Interpretation of velocity–time graphs

The diagram below shows a velocity–time graph. You must be able to interpret the graph and describe the motion it represents.

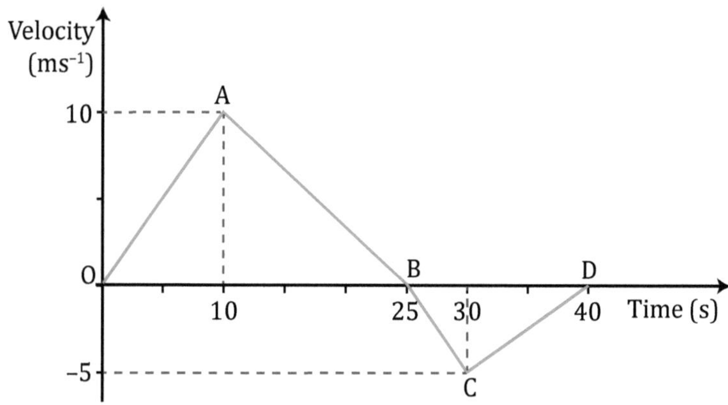

The graph has been divided into sections.

OA represents acceleration from 0 to 10 m s^{-1}.

Acceleration = the gradient of line OA = $\frac{10}{10}$ = 1 m s^{-2}.

AB represents a negative acceleration. The acceleration = $-\frac{10}{15}$ = −0.67 m s^{-2}. Note that the negative sign can be removed if it is described as a deceleration.

BC represents acceleration, but this time because the velocity is negative, it means that the motion is in the opposite direction.

The acceleration = $\frac{-5}{5}$ = −1 m s^{-2}. Notice that because the negative acceleration is in the same direction as the velocity, it represents acceleration rather than a deceleration.

The displacement from O to B is the area under the velocity–time graph between O and B.

Area = area of a triangle = $\frac{1}{2}$ × base × height = $\frac{1}{2}$ × 25 × 10 = 125 m

The displacement from B to D = $-\frac{1}{2}$ × 15 × 5 = −37.5 m

Note that because the area lies under the time axis it represents a negative displacement. This means that the object is moving back in the opposite direction so it now moving nearer to O.

The displacement from O to D = 125 − 37.5 = 87.5 m

The total distance travelled from O to D = 125 + 37.5 = 162.5 m

Active Learning

Produce a quiz for your maths friends. Draw a series of distance–time, displacement–time and velocity–time graphs and ask them to describe the motion shown. Then they can draw the graphs and test you.

7.5 Using the equations of motion (i.e. the *suvat* equations)

If a body moves with constant acceleration in a straight line, then the following formulae, called the equations of motion (or *suvat* equations) can be used. The meaning of the terms used in the equations is shown in the table:

$$v = u + at$$

$$s = ut + \frac{1}{2}at^2$$

$$v^2 = u^2 + 2as$$

$$s = \frac{1}{2}(u + v)t$$

s = displacement
u = initial velocity
v = final velocity
a = acceleration
t = time

Important note

You need to remember all the *suvat* equations as they will not be given in the exam.

Notice that there are four variables in each equation, so you would need to know three of them in order to find the value of the fourth. If there are two unknown variables then you need to find another equation connecting them so that they can be solved simultaneously.

Displacement, initial velocity, final velocity and acceleration are all vector quantities, which means they have both a magnitude (i.e., size) and a direction.

Normally the direction is taken as positive from left to right, so for example, a ball moving from right to left would have a negative velocity.

Note that in some questions the velocities in the above equations can be replaced by speeds.

The equations of motion can only be used when the motion is under constant acceleration.

Step by STEP

In the following question you are asked to derive one of the equations of motion using the velocity–time graph shown. This is an unstructured question as there are no steps to guide you how to arrive at the final formula. You have to work out the steps you need to take.

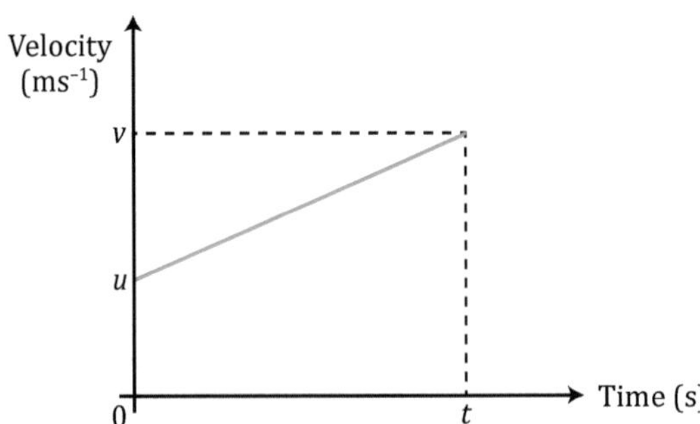

The graph above represents the motion of an aircraft taking off from a runway in a straight line. The aircraft accelerates with a constant acceleration of a m s^{-2} from a velocity of u m s^{-1} to a velocity of v m s^{-1} in time t seconds.

Use the above graph to prove the formula $v^2 = u^2 + 2as$ where s is the distance travelled in metres.

Steps to take

1 Look at the question carefully. You have a velocity–time graph. Think about the features of this graph. The gradient of the line is the acceleration and the distance travelled in time t is the area under the graph. Notice that the shape of the area is that of a trapezium.

2 Create equations using the letters on the graphs for the gradient (i.e. the acceleration) and the area under the graph (i.e. the distance travelled).

3 Notice that in the formula $v^2 = u^2 + 2as$ there is no t. This will mean if you obtain an equation which includes t, you will need to eliminate it using another equation.

Now follow the solution below.

Answer

The area under a velocity–time graph represents the distance travelled.

Area of trapezium $= \frac{1}{2}(u + v)t$

This is the distance travelled, s, so we can write

$$s = \frac{1}{2}(u + v)t$$

Notice that the formula you are asked to prove does not have t in it and also notice that it has a in it.

The gradient of the line represents the acceleration, so we can write

$$a = \frac{v - u}{t} \text{ and rearranging for } t \text{ gives } t = \frac{v - u}{a}$$

Substituting this for t in the equation $s = \frac{1}{2}(u + v)t$ gives

$$s = \frac{1}{2}(u + v)\left(\frac{v - u}{a}\right)$$

Multiplying both sides of the equation by $2a$ we obtain

$$2as = (u + v)(v - u)$$

$$2as = uv - u^2 + v^2 - uv$$

$$2as = -u^2 + v^2$$

Rearranging this to give the required format gives

$$v^2 = u^2 + 2as$$

Active Learning

Use the velocity–time graph shown in this question to derive the following equations of motion:

$$v = u + at$$

$$s = \frac{1}{2}(u + v)t$$

$$s = ut + \frac{1}{2}at^2$$

Use the facts that the gradient of the line is equal to the acceleration, and the area under the line equals the distance travelled, to help you with these derivations.

Solving problems using the *suvat* equations

The equations of motion can be used to solve problems that involve constant acceleration as the following examples show.

Examples

1 A toy car is given an initial velocity of $0.25\ \mathrm{m\,s^{-1}}$. Due to resistance, the deceleration is $0.25\ \mathrm{m\,s^{-2}}$. Find the distance travelled before the toy comes to rest.

Note here that you are told that the deceleration is $0.25\ \mathrm{m\,s^{-2}}$. This is an acceleration of $-0.25\ \mathrm{m\,s^{-2}}$ when used in the equations of motion.

Answer

1 First list the letters and their values when they are known.

$u = 0.25\ \mathrm{m\,s^{-1}}$, $a = -0.25\ \mathrm{m\,s^{-2}}$, $v = 0\ \mathrm{m\,s^{-1}}$, $s = ?$

Note you need to use one of the equations where you know the all the values of the letters except the one you need to find.

Using $v^2 = u^2 + 2as$ we have

$$0^2 = 0.25^2 + 2(-0.25)s$$

$$-0.0625 = -0.5\,s$$

$$s = 0.125\ \mathrm{m}$$

2 A particle is given an initial velocity of $4\ \mathrm{m\,s^{-1}}$ and is subject to a constant deceleration which brings the particle to rest in 5 seconds. Find the distance travelled by the particle.

Answer

2 $u = 4\ \mathrm{m\,s^{-1}}$, $v = 0\ \mathrm{m\,s^{-1}}$, $t = 5\ \mathrm{s}$, $s = ?$

Using $s = \frac{1}{2}(u + v)t$

$$= \frac{1}{2}(4 + 0)5$$

$$= 10\ \mathrm{m}$$

3 A particle, moves in a straight line and has its speed measured at points A and B. At point A, its speed is $20\ \mathrm{m\,s^{-1}}$ and at point B its speed is $32\ \mathrm{m\,s^{-1}}$. The distance between points A and B is 120 m.

(a) Show that the acceleration of the particle is $2.6\ \mathrm{m\,s^{-2}}$.

(b) Find the time for the particle to travel from A to B.

(c) Find the speed of the particle 20 s after passing point A.

(d) Calculate the distance from A 30 s after it passes A.

Answer

3 (a) $u = 20\ \mathrm{m\,s^{-1}}$, $v = 32\ \mathrm{m\,s^{-1}}$, $s = 120\ \mathrm{m}$, $a = ?$

Using $v^2 = u^2 + 2as$ we obtain

$$32^2 = 20^2 + 2a \times 120$$

Solving gives $a = 2.6\ \mathrm{m\,s^{-2}}$

(b) Using $v = u + at$ we obtain

$$32 = 20 + 2.6t$$

Solving we obtain $t = 4.62$ s

(c) $u = 20\ \mathrm{m\,s^{-1}}$, $a = 2.6\ \mathrm{m\,s^{-2}}$, $t = 20$ s, $v = ?$

Using $v = u + at$ we obtain

$$v = 20 + 2.6 \times 20$$

Solving we obtain $v = 72\ \mathrm{m\,s^{-1}}$

(d) $u = 20\ \mathrm{m\,s^{-1}}$, $a = 2.6\ \mathrm{m\,s^{-2}}$, $t = 30$ s, $s = ?$

Using $s = ut + \frac{1}{2}at^2$ we obtain

$$s = 20 \times 30 + \frac{1}{2} \times 2.6 \times 30^2$$

$$s = 1770\ \mathrm{m}$$

Step by STEP

A car is travelling along a straight horizontal road. There are three stages to its motion:

- During the first stage of the motion, it accelerates uniformly from rest with an acceleration of $1\ \mathrm{m\,s^{-2}}$ for 10 s.
- During the second stage of the motion, the car travels at constant velocity for 15 s.
- During the third stage of the motion, the car decelerates uniformly to rest in 5 s.

(a) Sketch a velocity–time graph that shows the three stages of the motion.

(b) Calculate the total distance travelled during the three stages of motion.

Steps to take

1 Always look at the entire question before starting. You can see that you are asked to sketch a velocity–time graph and then use it to calculate the distance. All the times are given which go on the horizontal axis but the constant velocity value is not given.

2 Find the constant velocity using a suitable *suvat* equation. Notice you know the starting velocity, acceleration and time over which it takes place.

3 Draw the graph.

4 Use the area under the graph to find the total distance travelled.

Answer

(a) $u = 0\ \mathrm{m\,s^{-1}}$, $a = 1\ \mathrm{m\,s^{-2}}$, $t = 10$ s, $v = ?$

Using $v = u + at$

$$v = 0 + 1 \times 10 = 10\ \mathrm{m\,s^{-1}}$$

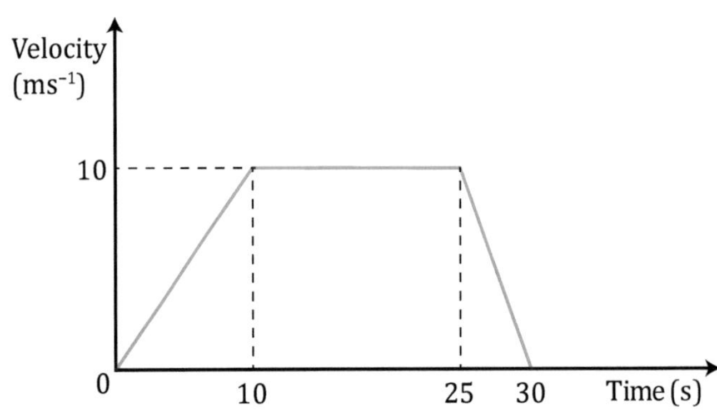

The formula for the area of a trapezium is used here which you will need to remember. If you cannot remember it, you can work out the areas of the two triangles and the rectangle and add them together.

(b) Total distance travelled = area under the velocity–time graph

$$= \frac{1}{2}(30 + 15) \times 10$$

$$= 225\text{ m}$$

---

4 A car is travelling along a straight road ABC with uniform acceleration a m s^{-2}. The distance AB is 95 m. The time taken by the car to travel from A to B is 5 s and the time taken to travel from B to C is 2 s. At A the speed of the car is u m s^{-1} and at C, its speed is 29.8 m s^{-1}. Find the value of a and the value of u. [7]

Answer

4

You need to consider two journeys; one from A to C and the other from A to B. If you considered a journey from B to C you would need to introduce another variable for velocity at B which would complicate things because there would now be three unknowns.

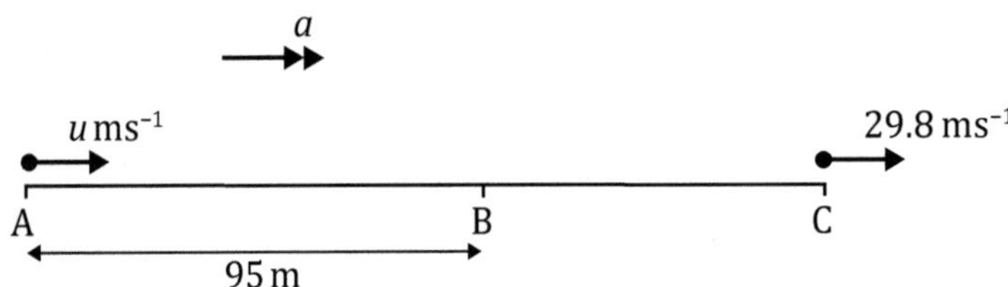

For the journey from A to C we have

$v = 29.8$ m s^{-1}, $t = 5 + 2 = 7$ s

Hence using $v = u + at$ gives

$$29.8 = u + 7a \quad (1)$$

Many of the questions in this topic involve the creation of two equations involving two unknowns which need to be solved simultaneously. Make sure you can solve simultaneous equations and check the answers you obtain.

For the journey from A to B we have

$s = 95$ m $t = 5$ s

Hence using $s = ut + \frac{1}{2}at^2$ gives

$$95 = 5u + 12.5a \quad (2)$$

Solving equations (1) and (2) simultaneously, we obtain

acceleration, $a = 2.4$ m s^{-2} and initial velocity, $u = 13$ m s^{-1}

5 A lorry travels along a straight road from point A to point B. The lorry starts from rest and accelerates uniformly for 25 s until it reaches a steady speed of 20 m s^{-1}. It travels at this steady speed for T s until it starts to decelerate uniformly for 10 s until it reaches point B. It passes point B with a speed of 10 m s^{-1} and the distance between points A and B is 12 km.

(a) Sketch a velocity–time graph for the journey between A and B. [3]

(b) Find the total time for the journey from A to B. [4]

(c) Describe two modelling assumptions you have made in your answers. [2]

Answer

5 (a)

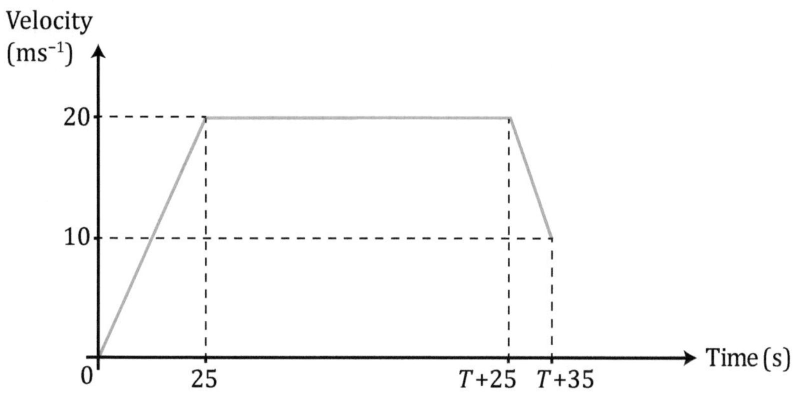

Notice the way the times are cumulative on the time axis.

(b) Distance travelled = area under the velocity–time graph

$$= \frac{1}{2} \times 25 \times 20 + 20T + \frac{1}{2} \times (20 + 10) \times 10$$

$$= 250 + 20T + 150$$

$$= 400 + 20T$$

Now distance travelled = 12 km = 12 000 m

$$12\,000 = 400 + 20T$$

$$11\,600 = 20T$$

$$T = 580 \text{ s}$$

(c) The lorry is modelled as a particle so it has a mass but no dimensions.

Frictional forces are assumed to be negligible, so the acceleration stays constant which means the equations of motion can be used.

7.6 Applying the *suvat* equations to vertical motion under gravity

Objects travelling in a vertical direction experience a constant acceleration of 9.8 m s^{-2} acting towards the centre of the Earth. When travelling upwards, this acceleration opposes the motion and is −9.8 m s^{-2}. When travelling downwards, this acceleration accelerates the particle and is 9.8 m s^{-2}. As displacements, velocities and accelerations can act in different directions, you have to decide which direction you intend to take as the positive direction.

The acceleration due to gravity does vary with distance to the centre of the Earth but as a modelling assumption we assume it remains constant.

Remember to say which direction you are taking as positive in your answers.

Examples

1 A ball is thrown vertically down a well with a velocity of 5 m s^{-1}. The ball takes 6 seconds to reach the bottom of the well. Take the acceleration due to gravity, g, as 9.8 m s^{-2}.

(a) Find the velocity of the ball after 6 seconds.

(b) Find the distance travelled by the ball when it reaches the bottom of the well.

Answer

Write down all the letters with values that are known and also write the letter of the quantity you wish to find with a question mark. Then you need to choose the equation from the list. It is a good idea to write down the list at the start as it will help you to remember them.

1 (a) Taking downwards as the positive direction, we have

$u = 5\text{ m s}^{-1}$, $a = g = 9.8\text{ m s}^{-2}$, $t = 6\text{ s}$, $v = ?$

Using $v = u + at$

$v = 5 + 9.8 \times 6 = 63.8\text{ m s}^{-1}$

(b) Using $s = ut + \frac{1}{2}at^2$ we have

$s = 5 \times 6 + \frac{1}{2} \times 9.8 \times 6^2 = 206.4\text{ m}$

2 A ball is thrown vertically upwards with a velocity of 10 m s^{-1}. Taking the value of the acceleration due to gravity, g, as 9.8 m s^{-2}.

(a) Find the maximum height reached by the ball.

(b) Find the time taken for the ball to reach its maximum height.

Answer

When the ball reaches its maximum height, its velocity is zero.

2 (a) Taking upwards as the positive direction, we have

$u = 10\text{ m s}^{-1}$, $v = 0\text{ m s}^{-1}$, $a = g = -9.8\text{ m s}^{-2}$, $s = ?$

Using $v^2 = u^2 + 2as$ we obtain

$0^2 = 10^2 + 2 \times (-9.8) \times s$ giving

$s = 5.10\text{ m}$

If upwards is taken as positive, then g is acting in the opposite direction so it has a negative value (i.e. −9.8).

(b) Using $v = u + at$ we obtain

$0 = 10 + (-9.8)t$ giving

$t = 1.02\text{ s}$

3 A stone is thrown vertically **downwards** from the top of a cliff with an initial velocity of 1 m s^{-1} and hits the sea 2.5 seconds later.

(a) Find the speed with which the stone hits the sea. [3]

(b) Calculate the height of the cliff. [3]

Answer

3 (a) Taking downwards as the positive direction, we have

$u = 1\ \text{m s}^{-1}$, $t = 2.5$ s, $a = g = 9.8\ \text{m s}^{-2}$, $v = ?$

Using $v = u + at$ we have

$v = 1 + 9.8 \times 2.5 = 25.5\ \text{m s}^{-1}$

(b) Using $s = ut + \frac{1}{2}at^2$

$= 1 \times 2.5 + \frac{1}{2} \times 9.8 \times 2.5^2$

$s = 33.125$ m

4 A stone is thrown vertically upwards with a speed of 14·7 m s^{-1} from a point A which is 49 m above the ground.

(a) Find the time taken for the stone to reach the ground. [3]

(b) Calculate the speed of the stone when it hits the ground. [3]

Answer

4 (a) Taking upwards as the positive direction, we have:

$u = 14.7\ \text{m s}^{-1}$, $s = -49$ m, $a = g = -9.8\ \text{m s}^{-2}$, $t = ?$

Using $s = ut + \frac{1}{2}at^2$ we obtain

$-49 = 14.7t + \frac{1}{2} \times (-9.8)t^2$

$-49 = 14.7t - 4.9t^2$

$0 = t^2 - 3t - 10$

$0 = (t - 5)(t + 2)$

Giving $t = 5$, or -2 (you cannot have a negative time so -2 is ignored).

Hence time taken = 5 s

(b) Using $v = u + at$ we have

$v = 14.7 - 9.8 \times 5$

$= -34.3\ \text{m s}^{-1}$

Hence, speed = 34.3 m s^{-1}

Notice we use $s = -49$. This is because the displacement is 49 m below the point of projection and therefore represents a negative displacement.

It looks as though you may need to use the quadratic formula to solve this quadratic equation. However, it is always worth checking to see if the coefficient of x^2 (4.9 in this case) divides exactly into the other numbers. On doing this here, you end up with a quadratic that can be easily factorised.

The negative sign tells us that this velocity is downwards (i.e. opposite to the direction of the initial velocity).

Speed is a scalar quantity so it only has size. We therefore remove the negative sign.

5 A boy throws a ball vertically upwards from a point A with an initial speed of 18.2 m s^{-1}.

(a) Find the greatest height above A reached by the ball. [3]

(b) Calculate the time taken for the ball to return to point A. [3]

(c) Find the speed of the ball 2.5 s after it was thrown. State clearly the direction of motion of the ball at this time. [3]

Note that at the maximum height reached, the final velocity, v, is zero.

Answer

5 (a) Taking the upward direction for the velocity as positive, we have

$u = 18.2\ \text{m s}^{-1}$, $a = g = -9.8\ \text{m s}^{-2}$, $v = 0\ \text{m s}^{-1}$, $s = ?$

Using $v^2 = u^2 + 2as$ we have

$$0^2 = 18.2^2 + 2 \times (-9.8) \times s$$

Rearranging and solving we have $s = 16.9$ m

As the ball returns to its original starting point, the displacement is zero.

(b) Taking the upward direction for the velocity as positive, we have

$u = 18.2\ \text{m s}^{-1}$, $a = g = -9.8\ \text{m s}^{-2}$, $s = 0$ m, $t = ?$

$$s = ut + \frac{1}{2}at^2$$

$$0 = 18.2t + \frac{1}{2} \times (-9.8) \times t^2$$

$$0 = t(18.2 - 4.9t)$$

Solving, gives $t = 0, 3.7$ s

Hence $t = 3.7$ s

$t = 0$ is ignored as this is the time when the ball first starts its journey. Hence $t = 3.7$ s.

As we have taken upward velocities as positive, the negative sign here means this velocity is downwards.

(c) Taking the upward direction for the velocity as positive, we have

$u = 18.2\ \text{m s}^{-1}$, $a = g = -9.8\ \text{m s}^{-2}$, $t = 2.5$ s, $v = ?$

$$v = u + at$$

$$= 18.2 - 9.8 \times 2.5 = -6.3\ \text{m s}^{-1}$$

Hence, the ball is moving downwards with a speed of 6.3 m s^{-1}

Active Learning

A stone is thrown vertically upwards. It arrives back at the same point from which it was thrown. Prove that the time to reach the greatest height is the same as the time to fall from this greatest height. What modelling assumptions did you make to obtain your answer?

Sketching velocity–time graphs where you have to use the equations of motion (*suvat* equations)

In some examination questions you will be asked to draw a velocity–time graph for motion described in a question. In some cases, you will need to add some information, which you will need to calculate first in order to draw your graph.

Examples

1 A particle, moves in a straight line and has its speed measured at points A and B. At point A, its speed is 20 m s^{-1} and at point B its speed is 32 m s^{-1}. The distance between points A and B is 120 m.

(a) Show that the acceleration of the particle is 2.6 m s^{-2}.

(b) Find the time for the particle to travel from A to B.

(c) Find the speed of the particle 20 s after passing point A.

(d) Calculate the distance from A 30 s after it passes A.

(e) Sketch a velocity–time graph for the journey from A to B.

Answer

1 (a) $u = 20\text{ m s}^{-1}$, $v = 32\text{ m s}^{-1}$, $s = 120\text{ m}$, $a = ?$

Using $v^2 = u^2 + 2as$ we obtain

$$32^2 = 20^2 + 2a \times 120$$

Solving gives $a = 2.6\text{ m s}^{-2}$

(b) Using $v = u + at$ we obtain

$$32 = 20 + 2.6t$$

Solving we obtain $t = 4.62$ s

(c) $u = 20\text{ m s}^{-1}$, $a = 2.6\text{ m s}^{-2}$, $t = 20\text{ s}$, $v = ?$

Using $v = u + at$ we obtain

$$v = 20 + 2.6 \times 20$$

Solving we obtain $v = 72\text{ m s}^{-1}$

(d) $u = 20\text{ m s}^{-1}$, $a = 2.6\text{ m s}^{-2}$, $t = 30\text{ s}$, $s = ?$

Using $s = ut + \frac{1}{2}at^2$ we obtain

$$s = 20 \times 30 + \frac{1}{2} \times 2.6 \times 30^2$$

$$s = 1770\text{ m}$$

(e)

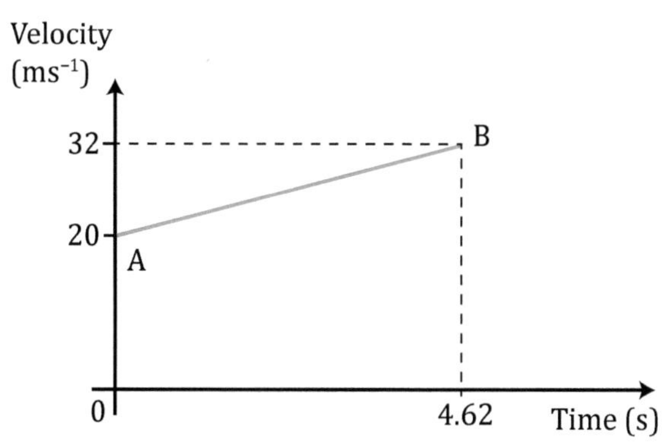

You should show any values you have calculated on your graph.

2 The points A, B and C lie, in that order, on a straight horizontal road. A car travels on the road with constant acceleration a m s^{-2}. When the car is at A, its speed is u m s^{-1}. The distance AB is 10 m and the car takes 2 s to travel from A to B. The car takes 7 s to travel from A to C and its speed at C is 17 m s^{-1}.

(a) Find the value of u and the value of a. [7]

(b) Draw a velocity–time graph for the motion of the car between A and C. [2]

(c) Calculate the distance AC. [2]

Lay out all the information given in the question in a clearly labelled diagram. Once drawn, study the diagram carefully to see what is known and what isn't.

Answer

2 (a)

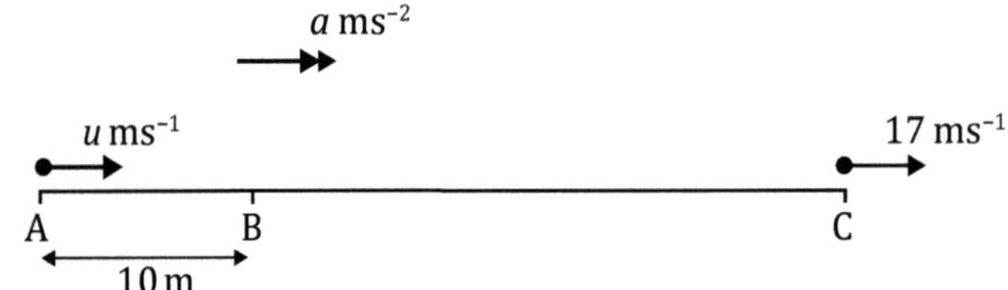

Go by the marks for each section of a question to see how much work each part involves; 7 marks is quite high so the solution is likely to involve quite a bit of work.

For A to B, $s = 10$ m, $t = 2$ s

Using $s = ut + \frac{1}{2}at^2$

$10 = 2u + \frac{1}{2}at^2$

$10 = 2u + \frac{1}{2} \times a \times 2^2$

$10 = 2u + 2a$ (1)

For A to C, $v = 17$ m s^{-1}, $t = 7$ s

Using $v = u + at$

$17 = u + 7a$ (2)

Solving equations (1) and (2) simultaneously, we obtain

$u = 3$ m s^{-1} and $a = 2$ m s^{-2}

You can put the value of u which you have just calculated, on the graph.

(b)

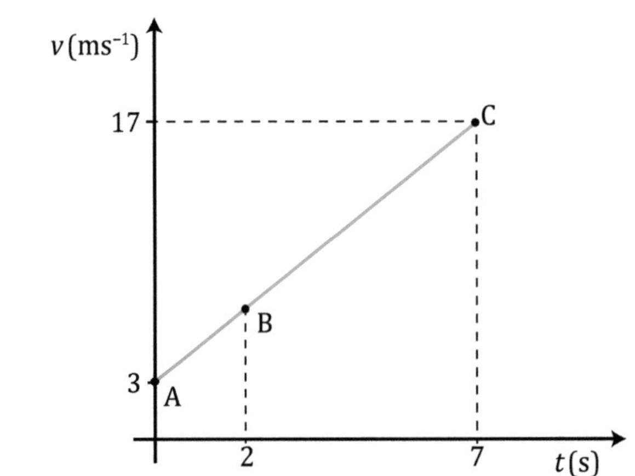

(c) The distance AC is the area under the velocity–time graph between A and C.

Area $= \frac{1}{2}(3 + 17) \times 7 = 70$

Distance $AC = 70$ m

The formula for the area of a trapezium is used here.

7.7 Using calculus in kinematics

Calculus (i.e. differentiation and integration) can be used to solve problems in kinematics when the acceleration is not constant. The equations of motion should be used for problems where there is a constant acceleration.

Displacement–time graphs when the velocity varies with time

The graph of displacement against time shown below is not linear so the graph does not represent constant velocity. The gradient of a displacement–time graph represents the velocity so the velocity at point P is lower than that at point Q. The gradient and hence the velocity will depend on the time chosen.

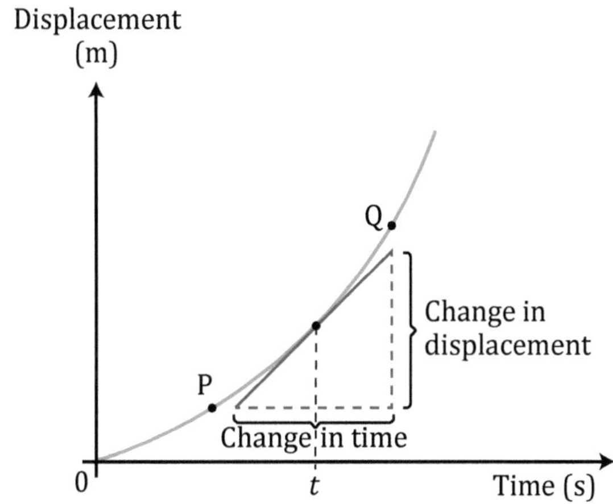

$$\text{velocity} = \frac{\text{change in dislacement}}{\text{change in time}} \quad \text{(i.e. the gradient)}$$

Finding the velocity from the displacement

The velocity at any time t is given by the gradient of the displacement–time graph.

Hence we have the important result $v = \frac{dr}{dt}$

To find the velocity numerically, the time at which the velocity is to found is substituted into the expression.

To find the velocity from the displacement, you differentiate the expression for the displacement with respect to time.

When the acceleration is not constant, we use the letter r instead of the letter s for the displacement.

Finding the acceleration from the velocity

The acceleration at any time t is given by the gradient of the velocity–time graph.

Hence we have the important result

$$a = \frac{dv}{dt} = \frac{d^2r}{dt^2}$$

Displacement (r) —Differentiate $\frac{dr}{dt}$→ Velocity (v) —Differentiate $\frac{dv}{dt}$→ Acceleration (a)

To find the acceleration from the velocity, you differentiate the expression for the velocity with respect to t $\left(\text{i.e. } a = \frac{dv}{dt}\right)$ and if you have an expression for the displacement you differentiate the displacement expression twice $\left(\text{i.e. } \frac{d^2r}{dt^2}\right)$.

Differentiating the expression for r with respect to t.

Remember that to differentiate you multiply by the index and then reduce the index by 1.

Differentiating the expression for v with respect to t.

To integrate a term you increase the index by one and divide by the new index.

Remember that when integrating you must remember to include the constant of integration.

Example

1 A particle moves along a straight line in such a way that the displacement r at time t is given by

$$r = 12t^3 - 6t^2 + 1$$

(a) Find an expression for the velocity, v, at time t.

(b) The particle is at rest at two times. Find these two times.

(c) Find an expression for the acceleration, a, at time t.

Answer

1 (a) $v = \frac{dr}{dt}$

$= 36t^2 - 12t$

(b) When at rest, $v = 0$

$$0 = 36t^2 - 12t$$

$$0 = 12t(3t - 1)$$

Solving gives $t = 0$ s or $\frac{1}{3}$ s

(c) $a = \frac{dv}{dt}$

$= 72t - 12$

Working backwards to find the velocity and the displacement

If you need to work backwards through the diagram shown in the previous section then you will need to reverse the effect of differentiating by integrating as the diagram below shows.

Displacement (r) ← Integrate, $r = \int v\, dt$ ← Velocity (v) ← Integrate, $v = \int a\, dt$ ← Acceleration (a)

Finding the velocity from the acceleration

To find the velocity from the acceleration, you integrate the expression for the velocity with respect to t.

Hence we have the important result

$$v = \int a\, dt$$

For example, if $a = 2 - t$ then $v = \int a\, dt = \int (2 - t)\, dt = 2t - \frac{t^2}{2} + c$

Notice the inclusion of the constant of integration, c. It is necessary to find the value of c and this is done by substituting known values for v and t into the expression for v.

For example, if it is known that the object starts from rest, we can say that when $t = 0$, $v = 0$ so

$$0 = 2(0) - \frac{(0)^2}{2} + c \text{ and solving gives } c = 0.$$

The constant of integration, c, is now substituted into the expression to give

$$v = 2t - \frac{t^2}{2}$$

Finding the displacement from the velocity

To find the displacement from the velocity, the velocity expression is integrated with respect to t.

Hence we have the important result

$$r = \int v\,\mathrm{d}t$$

Carrying on with our example, to find the displacement we need to integrate the velocity expression, $v = 2t - \frac{t^2}{2}$ with respect to t.

Hence we have $$r = \int v\,\mathrm{d}t = \int\left(2t - \frac{t^2}{2}\right)\mathrm{d}t = t^2 - \frac{t^3}{6} + c$$

Again we need to substitute known values of r and t into the expression to find the value of the constant c.

If we knew that when $t = 0$, $r = 0$, we would have

$$0 = (0)^2 - \frac{(0)^3}{3} + c, \quad \text{giving } c = 0$$

The constant of integration, c, is now substituted into the expression to give

$$r = t^2 - \frac{t^3}{3}$$

Examples

1 A particle starts from rest and travels from O along the positive x-axis in a straight line. The particle has acceleration, a m s^{-2}, given by

$$a = 4t - 3t^2$$

(a) Find an expression for the velocity, v m s^{-1}, in time t.

(b) Find an expression for the displacement, r m, in time t.

Answer

1 (a) $v = \int a\,\mathrm{d}t$

$= \int (4t - 3t^2)\,\mathrm{d}t$

$= \frac{4t^2}{2} - \frac{3t^3}{3} + c$

$= 2t^2 - t^3 + c$

As the particle starts from rest, when $t = 0$, $v = 0$

Hence we have $0 = 2(0)^2 - (0)^3 + c$

We need to find the value of the constant of integration, c, by substituting a pair of known values for v and t into this equation.

Solving gives $c = 0$

Hence the expression for the velocity is

$$v = 2t^2 - t^3$$

(b) $r = \int v\,dt$

$= \int (2t^2 - t^3)\,dt$

$= \frac{2t^3}{3} - \frac{t^4}{4} + c$

When $t = 0$, $r = 0$

$$0 = \frac{2(0)^3}{3} - \frac{(0)^4}{4} + c \quad \text{giving } c = 0$$

Hence the expression for the displacement is

$$r = \frac{2t^3}{3} - \frac{t^4}{4}$$

---

2 A car accelerates from rest and at a time t seconds, its acceleration is given by $a = 5 - 0.1t\,\text{m s}^{-2}$

(a) Find the velocity after 10 seconds.

(b) Find the time when the acceleration is zero.

(c) Find the distance travelled in the first 50 seconds, giving your answer to the nearest metre.

Answer

2 (a) $v = \int a\,dt$

$= \int (5 - 0.1t)\,dt$

$= 5t - \frac{0.1t^2}{2} + c$

As the car starts from rest, we know when $t = 0$, $v = 0$

So $\quad 0 = 5(0) - \frac{0.1(0)^2}{2} + c$

Solving gives $c = 0$

Hence we have $\quad v = 5t - \frac{0.1t^2}{2}$

$$= 5(10) - \frac{0.1(10)^2}{2} = 45\,\text{m s}^{-1}$$

(b) As $a = 0$, $0 = 5 - 0.1t$

Solving gives $t = 50$ s

(c) $r = \int v\,dt$

$= \int \left(5t - \frac{0.1t^2}{2}\right) dt$

$= \frac{5t^2}{2} - \frac{0.1t^3}{6} + c$

As the car starts from rest when $t = 0$, $r = 0$

Hence $c = 0$

So we have $r = \frac{5t^2}{2} - \frac{0.1t^3}{6}$

when $t = 50$, $r = \frac{5(50)^2}{2} - \frac{0.1(50)^3}{6}$

$= 6250 - 2083$

$= 4167$ m (nearest metre)

---

3 The acceleration of a moving particle is given by

$(3 + 2t)\ \text{m s}^{-2}$.

The initial speed of the particle is $10\ \text{m s}^{-1}$.

Find the speed after 2 seconds.

Answer

3 $a = 3 + 2t$

$v = \int a\,dt$

$= \int (3 + 2t)dt$

$= 3t + \frac{2t^2}{2} + c$

$= 3t + t^2 + c$

When $t = 0$, $v = 10$

so $10 = 3(0) + (0)^2 + c$ giving $c = 10$

Hence $v = 3t + t^2 + 10$

When $t = 2$, $v = 3(2) + (2)^2 + 10 = 20\ \text{m s}^{-1}$

BOOST Grade

Do not assume that the constant of integration, c, is zero in all cases. Here is an example where the non-zero value of c has to be found.

Using definite integration to find the distance/displacement travelled in a time interval (i.e. between two times)

To find the distance/displacement travelled during a time interval (i.e. between two times t_1 and t_2) you integrate the velocity expression between the limits of the two times.

Hence

$r = \int_{t_1}^{t_2} v\,dt$ where t_1 and t_2 are the two times and where t_2 is the later time.

Example

1 A particle moves in a straight line such that t seconds after leaving point O its velocity in m s^{-1} is given by the expression

$$v = 3t^2 - 2t$$

Find the distance travelled in the 3rd second.

The larger time will be the top limit.

Remember to remove the integration sign after integration and use the square brackets with the limits on them.

Now replace t with the top limit and then with the bottom limit and separate the two brackets with a minus sign.

BOOST Grade

Remember to consider direction when performing calculations using vector quantities. You have to decide which direction you are taking as positive and state this in your answer.

The *suvat* equations can only be used if there is constant acceleration. You have to use calculus for questions involving variable acceleration.

Remember to label the axes when drawing graphs with the name of the variables and units.

Answer

1 Note that 0 to 1 is the 1st second, 1 to 2 is the 2nd second so 2 to 3 is the 3rd second.

Hence we integrate using the limits of 3 and 2, with 3 being the upper limit.

$$r = \int_{t_1}^{t_2} v\,dt$$

$$= \int_2^3 v\,dt$$

$$= \int_2^3 (3t^2 - 2t)\,dt$$

$$= \left[\frac{3t^3}{3} - \frac{2t^2}{2}\right]_2^3$$

$$= \left[t^3 - t^2\right]_2^3$$

$$= \left[(3^3 - 3^2) - (2^3 - 2^2)\right]$$

$$= 18 - 4$$

$$= 14\text{ m}$$

Test yourself

1 A car, initially at rest, accelerates with a constant acceleration of 0.9 m s^{-2}. Calculate:
(a) The speed of the car after 10 seconds.
(b) The distance travelled in this time.

2 A stone is projected vertically upwards with a velocity of 20 m s^{-1} from point A.
(a) Calculate the greatest height from point A reached by the stone.
(b) Calculate the time from when the stone is projected to when it returns to point A.

3 A stone is thrown vertically downwards from the top of a cliff with a velocity of 0.8 m s^{-1} and hits the sea 3.5 seconds later.
(a) Calculate the speed with which the stone hits the sea.
(b) Calculate the height of the cliff.

4 A particle is projected vertically upwards with speed 10 m s^{-1}.
(a) Find the time in seconds for the particle to reach its greatest height.
(b) Find the maximum height reached by the particle.

5 A particle moves along a straight line. At time t seconds, its displacement, r metres from the origin is given by $r = 12t^3 + 9$.
(a) Find an expression for the velocity of the particle at time t.
(b) Find the acceleration at time $t = 2$ seconds.

6 A particle accelerates from rest. t seconds after starting its motion it has a velocity $v \text{ m s}^{-1}$ given by $v = 0.64t^3 - 0.36t^2$.
(a) Find an expression for the acceleration at time t.
(b) Find the distance travelled after 10 seconds.

7 A lorry accelerates from rest and at time t seconds, its acceleration is given by $a = 3 - 0.1t$ until $t = 30$ s.
(a) Find an expression in terms of t for the velocity of the lorry.
(b) Find the velocity of the lorry after 10 seconds.
(c) Explain what will happen to the lorry at $t = 30$ seconds.
(d) Find the distance travelled in the first 30 seconds.

8 A particle moves in a straight line and its velocity is $v \text{ m s}^{-1}$, t seconds after passing the origin O where v is given by $v = 6t^2 + 4$
Find the distance travelled between the times $t = 2$ s and $t = 5$ s.

9 A particle moves in a straight line and at time t seconds, it has velocity $v \text{ m s}^{-1}$, where

$$v = 6t^2 - 2t + 8$$

(a) (i) Find an expression for the acceleration of the particle at time t.
(ii) Find the acceleration of the particle when $t = 1$ second.
(b) When $t = 0$ s, the particle is at the origin. Find an expression for the displacement of the particle from the origin at time t.

10 A fighter jet lands on the deck of an aircraft carrier. The landing system used reduces the velocity of the aircraft from 153 km h^{-1} to 0 km h^{-1} in 2 seconds.
(a) Calculate the deceleration of the aircraft in m s^{-2}. [3]
(b) If the length of the deck of the aircraft carrier is 300 m, what fraction of the deck is used in landing? [2]

11 The graph below shows the velocity–time graph for the motion of a car travelling in a straight line over 55 seconds.

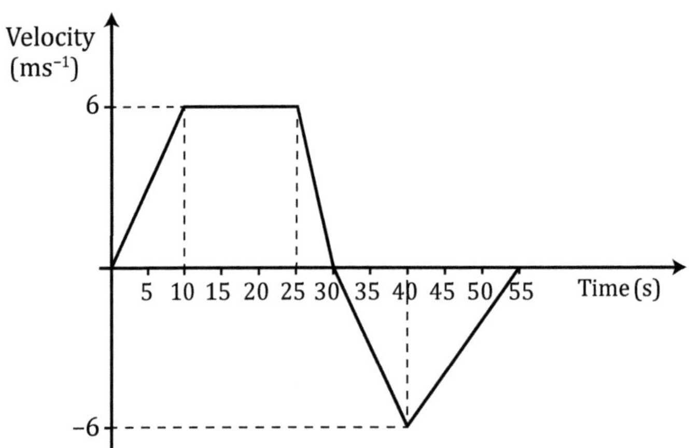

(a) Calculate the distance travelled by the car in the first 30 seconds. [2]
(b) Calculate the total distance travelled in the 55 seconds of the car's motion. [1]
(c) Find the displacement of the car from its initial position after 55 seconds. [2]
(d) Calculate the average speed of the car for the motion represented by the graph giving your answer to 2 significant figures. [2]

12 A stone is thrown vertically down a well with a velocity of $2\ \text{m s}^{-1}$. The stone hits the water 3 seconds later.
(a) Calculate the distance travelled by the stone before hitting the water. [2]
(b) Calculate the velocity with which the stone hits the water. [2]
(c) State two modelling assumptions you have made in order to arrive at your answers to parts (a) and (b). [2]

13 A speeding car travelling at a constant $14\ \text{m s}^{-1}$ along a straight road passes a parked police car. Two seconds after the speeding car passes, the police car sets off in pursuit of the speeding car. The police car has a constant acceleration of $4\ \text{m s}^{-2}$.
(a) Draw a velocity–time graph to show the motions of the two cars. [2]
(b) Using your velocity–time graph, or otherwise, find the time after being passed by the speeding car that the police car will draw level with the speeding car. Give your answer to 2 significant figures. [5]
(c) Calculate the distance travelled by the police car to draw level. Give your answer to 2 significant figures. [2]

14 A particle accelerates from rest so that t seconds after starting its velocity, in m s^{-1}, is given by the formula

$$v = t^3 - 2t^2 + t$$

(a) Find the times when the velocity of the particle is zero. [2]
(b) Find the acceleration at time t. [2]
(c) Find the distance travelled in the first second. [2]

15 A metal ball is released from rest at the surface of a thick liquid and at time t seconds its depth, r metres is given by $r = 3t^2 - 4t^3$
Find the greatest depth reached by the ball. [5]

Summary

Check you know the following facts and formulae:

Displacement/distance–time graphs

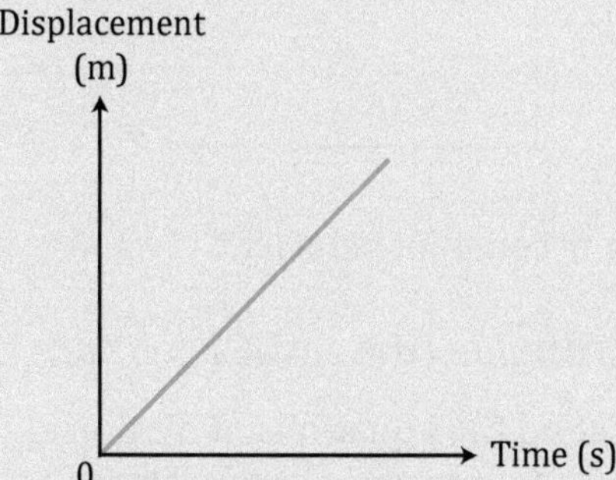

The gradient represents velocity/speed.

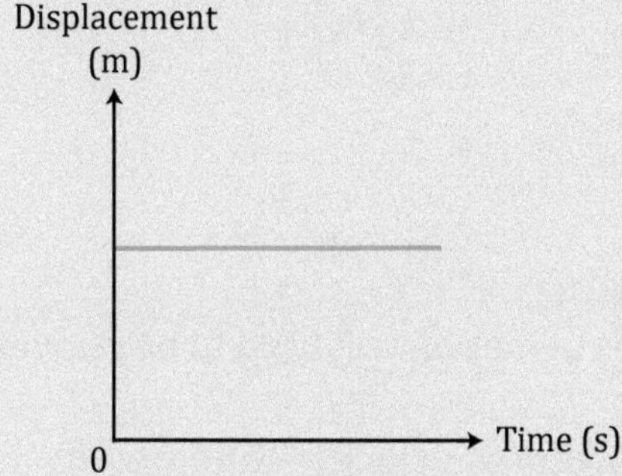

A horizontal line has zero gradient and represents a body at rest.

Velocity–time graphs

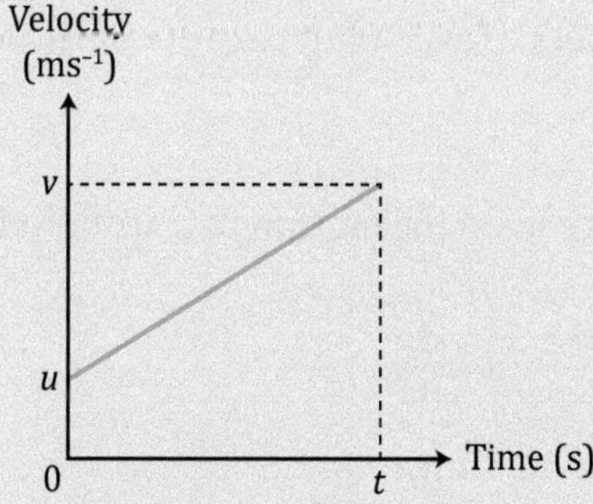

The gradient represents acceleration.

The area under the graph represents the distance/displacement.

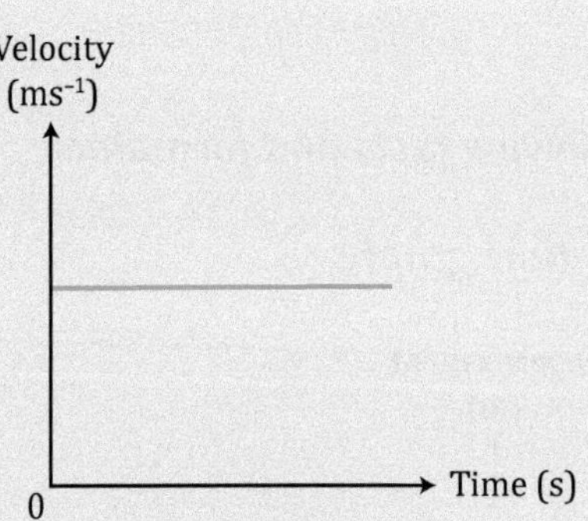

A horizontal line represents constant velocity.

Constant acceleration formulae (the suvat equations)

s = displacement or distance
u = initial velocity
v = final velocity
a = acceleration
t = time

The following formulae, called the equations of motion should only be used if it is known that the motion is under constant acceleration in a straight line.

$$v = u + at$$

$$s = ut + \frac{1}{2}at^2$$

$$v^2 = u^2 + 2as$$

$$s = \frac{1}{2}(u + v)t$$

Note that all the equations of motion will need to be remembered.

Sketching and interpretation of velocity–time graphs

When asked to sketch a velocity–time graph:

Use the equations of motion to find unknown quantities if needed.

Remember you are creating a sketch so you do not need a scale along each axis. You only need to include the important values or values asked for in the questions.

Remember to label both sets of axes with the title of the axis and unit.

Kinematics

The following formulae are used for motion in a straight line when the acceleration varies with time:

These kinematics formulae will be given on the formula sheet.

$$v = \frac{\mathrm{d}r}{\mathrm{d}t}$$

$$a = \frac{\mathrm{d}r}{\mathrm{d}t} = \frac{\mathrm{d}^2r}{\mathrm{d}t^2}$$

$$r = \int v\,\mathrm{d}t$$

$$v = \int a\,\mathrm{d}t$$

Active Learning

Take a picture of the graphs and *suvat* equations page with your phone. You need to remember all this material. Keep looking at the page to help memorise it.

8 Dynamics of a particle

Introduction

Dynamics is the branch of mechanics concerned with the motion of bodies under the action of forces. This topics looks at a set of laws called Newton's laws of motion and how forces can change the motion of particles.

This topic covers the following:

8.1 Newton's laws of motion

8.2 Types of force

8.3 Lifts accelerating, decelerating and travelling with constant velocity

8.4 The motion of particles connected by strings passing over pulleys or pegs

8.1 Newton's laws of motion

Force is a vector quantity as it has both magnitude and direction.

Unbalanced forces produce an acceleration. The unit of force is the newton (N) and 1 N is the force which causes a 1 kg mass to be accelerated at 1 m s^{-2}.

Newton's first law states that a particle will remain at rest or will continue to move with constant speed in a straight line unless acted upon by some external force. This means that if a particle is acted upon by unbalanced forces then its velocity will change.

Newton's second law states that a resultant force produces an acceleration, according to the formula

$$\text{force} = \text{mass} \times \text{acceleration} \quad \text{or} \quad F = m\,a$$

Newton's third law states that every action has an equal and opposite reaction. This means that if body A exerts a force on body B, then body B will exert an equal and opposite force on body A.

Using the formula $F = ma$ (Newton's second law) to solve problems

According to Newton's second law of motion, an unbalanced force will produce an acceleration. Take the following situation where we will assume that all the forces act only in the horizontal direction.

Always try to work out the direction of the acceleration and mark this on your diagram.

2N ← 3kg → 8N

In order to find the resultant force we resolve the forces in a certain direction. Normally you resolve the forces in the direction of the acceleration. As the resultant force will be to the right (because the larger of the two forces is to the right), the acceleration will be to the right so we would take the right as the positive direction.

There is a resultant force of 8 – 2 = 6 N to the right and this force will produce an acceleration to the right according to the equation, $F = ma$.

Putting values for the mass and resultant force into this equation gives

$$6 = 3a, \quad \text{giving} \quad a = 2\text{ m s}^{-2}.$$

Example

1 For each of the following situations, find the resultant force acting on the block and the acceleration.

(d)

4 N ← [2 kg] → 2 N
1 N ← → 3 N

Answer

1 (a) Resultant force = 10 − 8 = 2 N to the right.

Using $F = ma$

$2 = 2a$

Hence acceleration $a = 1\ \text{m s}^{-2}$ to the right.

(b) Resultant force = 6 − 2 = 4 N to the left.

Using $F = ma$

$4 = 2a$

Hence acceleration $a = 2\ \text{m s}^{-2}$ to the left.

(c) Resultant force = 5 + 8 − 3 = 10 N to the right.

Using $F = ma$

$10 = 2a$

Hence acceleration $a = 5\ \text{m s}^{-2}$ to the right.

(d) Resultant force = (2 + 3) − (4 + 1) = 0 N

Using $F = ma$

$0 = 2a$

Hence acceleration $a = 0\ \text{m s}^{-2}$.

Examples of Newton's laws of motion

Newton's first law

In the absence of any force (for example, in outer space away from any gravitational attraction of stars, planets, etc.) once a body is given a velocity it will continue with the velocity until it experiences a force. If there is an unbalanced force (called a resultant force), then this will always produce an acceleration.

Newton's second law

If there is an unbalanced force (i.e. resultant force), then this force will produce an acceleration and the acceleration will be in the same direction as the resultant force.

Newton's third law

Here a mass is hung from a string and the arrangement is stationary. The weight (i.e. the action force of $10g$) acts vertically down and the reaction force is the tension in the string which acts vertically up. Action and reaction are equal and opposite so the tension in the string is also $10g$ but upwards.

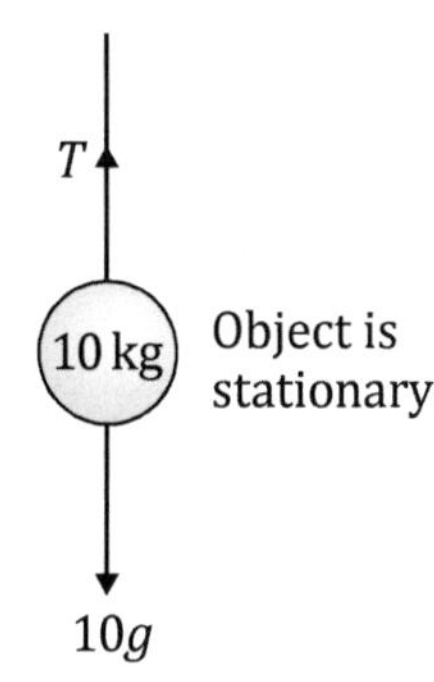

If the arrangement shown was accelerating upwards or downwards the tension would no longer be equal in magnitude to the weight. It would only be equal if the arrangement was at rest or travelling with constant velocity. You will learn more about these situations later.

8.2 Types of force

In this topic you will be looking at the effects of certain types of force on a particle. These forces include the following:

- Weight
- Friction
- Normal reaction
- Tension
- Thrust.

Weight is the attractive force between the Earth and the particle. It always acts in a downward direction towards the Earth's centre. Weight can be calculated by multiplying the mass of the particle in kg by the acceleration due to gravity, g, which has the value 9.8 m s^{-2}. Hence

$$\text{weight} = mg$$

Friction is the force which opposes movement if the surface on which the particle is placed is rough. If you are told that a surface is smooth, then the friction will be zero.

Normal reaction is the force a surface exerts on a particle perpendicular to the surface when there is contact between the two. For example, a particle on a level surface will exert a force (i.e. its weight) vertically down on the surface. According to Newton's third law, the surface will exert an equal but opposite force on the particle for the system to remain in equilibrium. This equal but opposite force is called the normal reaction.

Tension is a resisting force in a string which opposes any tendency for the string to extend. For example, if a particle is hung from a string, the weight of the particle acts down and the tension acts upward to maintain equilibrium.

Thrust is a resisting force provided by a spring. It always acts in a direction to oppose the force that is either compressing or extending the spring.

Example

1 A block of mass 5 kg rests on a smooth horizontal surface. The block is then subjected to a force of 10 N acting horizontally to the right.

(a) Draw a diagram showing all the forces acting on the block.

(b) Find the size of the normal reaction.

(c) Find the acceleration of the block.

Answer

1 (a)

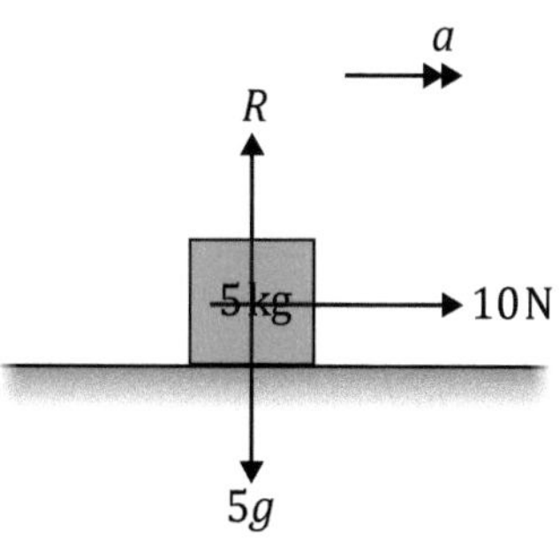

Always try to work out the direction of the acceleration and mark this on your diagram.

(b) As there is no vertical motion, the vertical forces must be equal.

Hence $R = 5g = 49$ N.

This is called resolving the forces vertically. As the vertical forces are in equilibrium we can equate the upward force with the downward force.

(c) Resultant force = 10 N to the right.

Using $F = ma$

$10 = 5a$

Hence acceleration $a = 2$ m s^{-2} to the right.

Always remember that acceleration and forces are vector quantities. You should always give their directions unless the question only asks for their magnitude.

8.3 Lifts accelerating, decelerating and travelling with constant velocity

This section deals with objects on the floor of lifts when the lift has an acceleration in either direction or when the lift is travelling at constant speed or at rest.

Newton's second law of motion can be applied to each of these situations where all the blocks inside the lift have a mass of m kg. In these situations, the reaction of the lift floor on the body is R upwards.

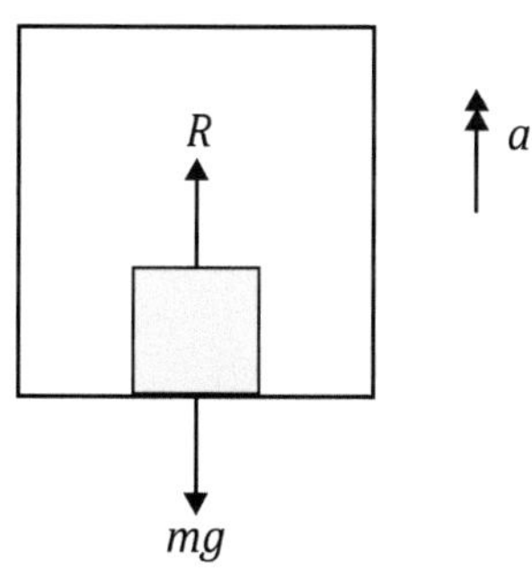

There is a resultant force acting in the direction of the acceleration (i.e. upwards).

This means that R is larger than mg.

Hence $ma = R - mg$

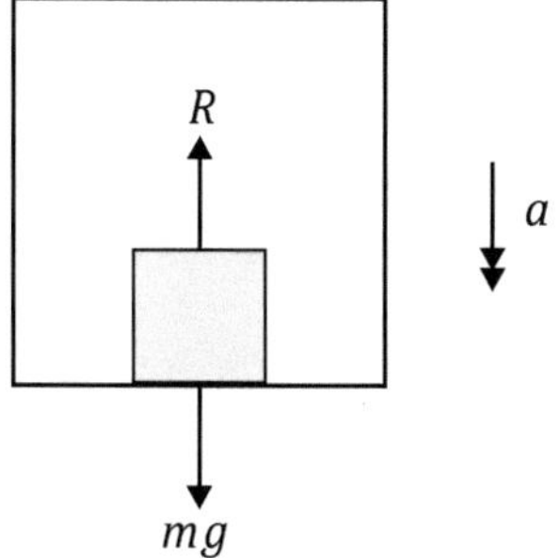

There is a resultant force acting in the direction of the acceleration (i.e. downwards).

This means that mg is larger than R.

Hence $ma = mg - R$

There is no resultant force acting because the acceleration is zero.

This means that R is equal to mg.

Hence $R = mg$

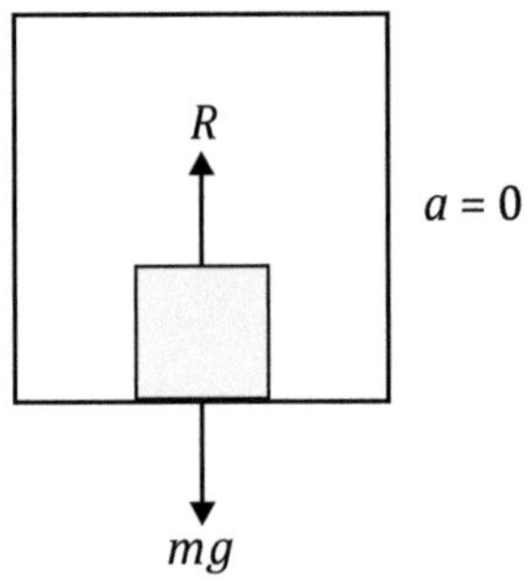

BOOST Grade

Always draw a diagram of the arrangement and mark on the diagram the direction of the acceleration. It is advisable to take the direction of the acceleration as the positive direction for forces.

Important note

In the above diagrams the actual direction of movement of the lift has not been included. It is only the direction of the acceleration that is important. For example, the first diagram could apply to either a lift moving upwards with an acceleration a or it could apply to a lift moving downwards which is decelerating with a deceleration a.

Active Learning

A friend says that it does not matter which way (up or down) a lift is travelling as the only thing that affects the tension in the cable or the normal reaction on the person standing in lift is the direction of the acceleration. Is your friend right or wrong? Explain your answer.

Examples

1 A box, of mass 30 kg, rests on the floor of a lift. Find the reaction of the floor of the lift on the crate when:

(a) the lift is moving up with acceleration 0.3 m s^{-2}

(b) the lift is moving down with acceleration 0.2 m s^{-2}

(c) the lift is moving up with constant speed.

Answer

1 (a)

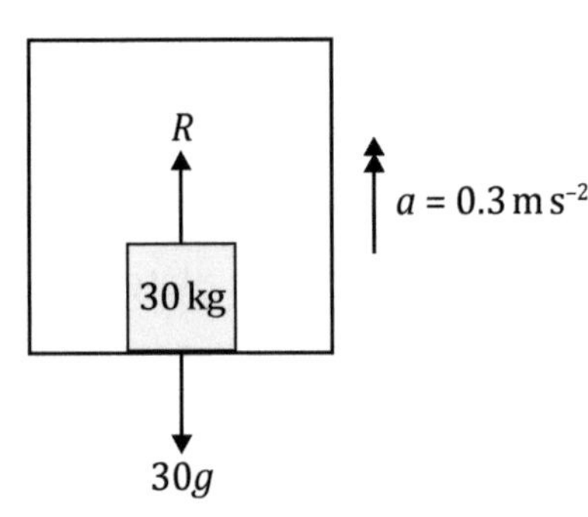

Applying Newton's second law to the box, we have

$$ma = R - mg$$

$$30 \times 0.3 = R - (30 \times 9.8)$$

giving $R = 303$ N

(b)

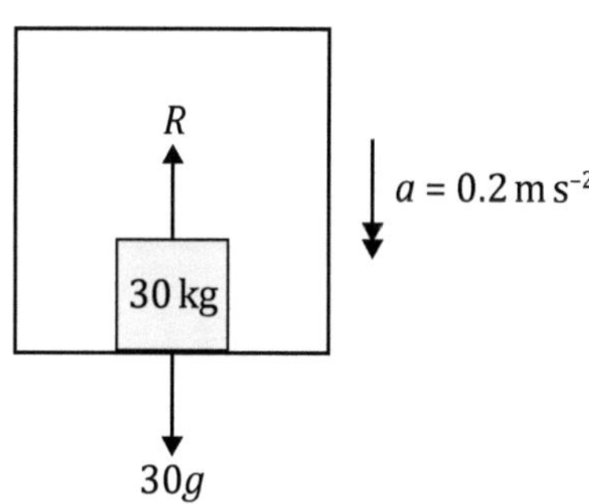

Applying Newton's second law to the box, we have

$$ma = mg - R$$

$$30 \times 0.2 = (30 \times 9.8) - R$$

giving $R = 288$ N

(c)

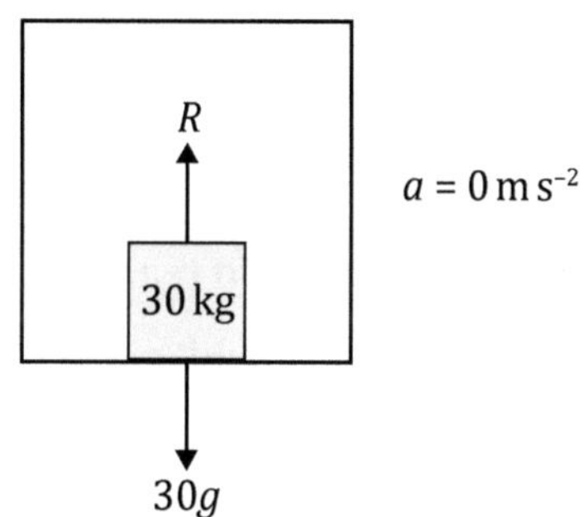

The lift is travelling at constant speed, so the acceleration is zero. The weight and reaction balance so there is no resultant force.

Applying Newton's second law to the box, we have

$$R = mg = 30 \times 9.8 = 294\text{ N}$$

---

2 A person, of mass 60 kg, is standing in a lift, which is of mass 540 kg. When the lift is accelerating upwards at a constant rate of a m s^{-2}, the tension in the lift cable is 6600 N.

(a) Calculate the value of a. [3]

(b) Find the reaction between the person and the floor of the lift. [3]

Answer

2 (a)

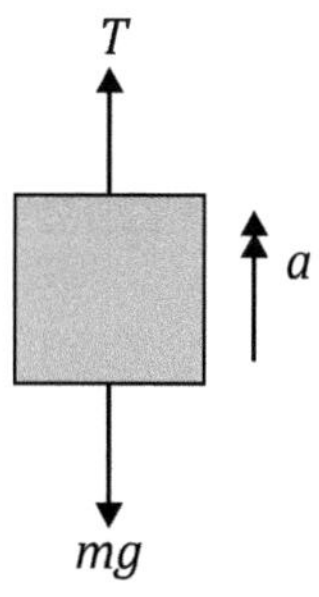

Applying Newton's second law to the lift, we have

$$ma = T - mg$$

The mass in this equation is the mass of the lift and the person added together.

Hence $600a = 6600 - (600 \times 9.8)$

giving $a = 1.2\ \text{m}\,\text{s}^{-2}$

(b)

The only forces acting directly on the person are the weight downwards and the reaction of the lift floor on the person.

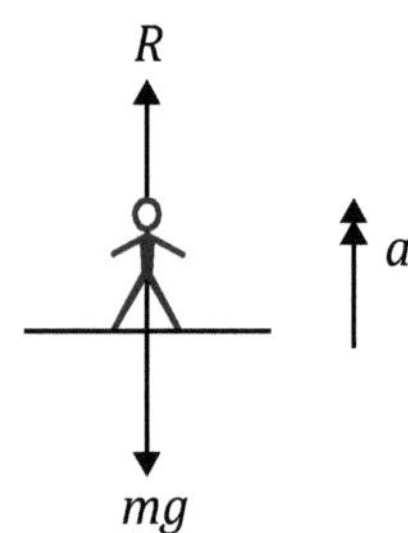

Applying Newton's second law to the person, we have

$$ma = R - mg$$

$$60 \times 1.2 = R - (60 \times 9.8)$$

giving $R = 660\ \text{N}$

---

3 A lift is moving upwards. It accelerates from rest with uniform acceleration $0.4\ \text{m}\,\text{s}^{-2}$ until it reaches a speed of $2\ \text{m}\,\text{s}^{-1}$. It then travels at this constant speed of $2\ \text{m}\,\text{s}^{-1}$ for 17 s before decelerating uniformly to rest in 8 s.

(a) Calculate the time taken for the lift to reach the speed of $2\ \text{m}\,\text{s}^{-1}$. [3]

(b) Sketch a velocity–time graph for the lift's journey. [3]

(c) Find the distance travelled by the lift during the journey. [3]

(d) A man, of mass 70 kg, is standing in the lift during its journey. Calculate the greatest value of the reaction exerted by the floor of the lift on the man during the journey. [4]

Answer

3 (a) Using $v = u + at$

with $v = 2\ \text{m}\,\text{s}^{-1}$, $u = 0\ \text{m}\,\text{s}^{-1}$, $a = 0.4\ \text{m}\,\text{s}^{-2}$ we obtain

$2 = 0 + (0.4 \times t)$, giving $t = 5$ s.

(b)

You could, alternatively, split the shape up and find the area of the two triangles and the rectangle and then add them together.

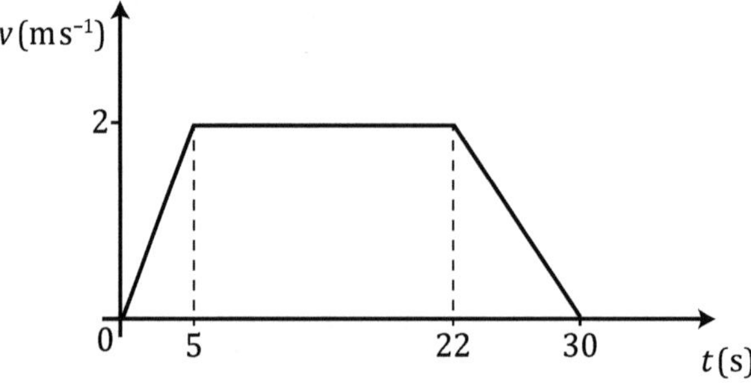

(c) Distance travelled = area under the velocity–time graph.

Area of a trapezium = $\frac{1}{2}$(sum of the two parallel sides) × distance between the parallel sides

$= \frac{1}{2}(30 + 17) \times 2 = 47\ \text{m}$

(d)

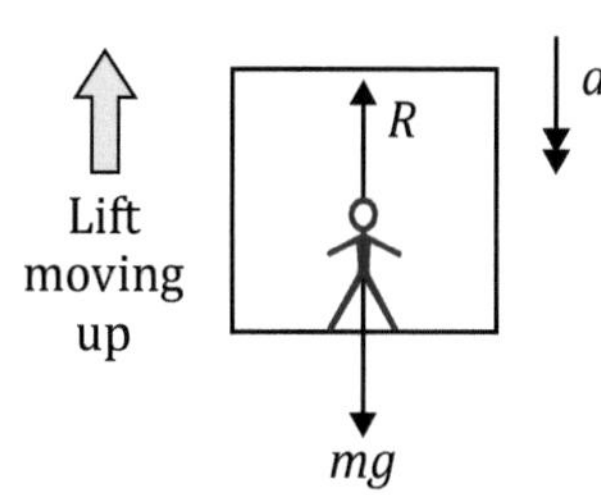

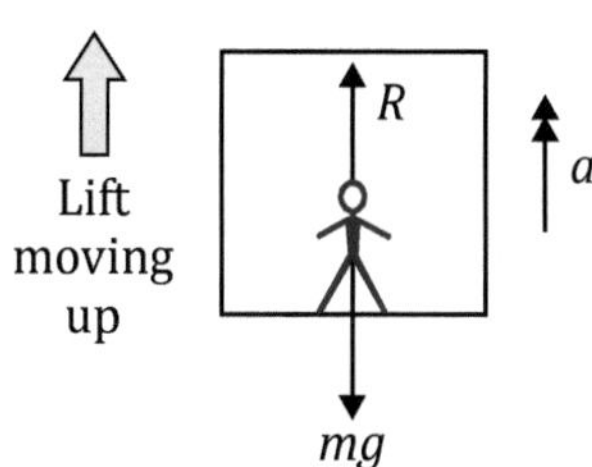

When the lift decelerates when travelling up, applying Newton's second law, we have $ma = mg - R$.

Hence $R = mg - ma = (70 \times 9.8) - (70 \times 0.25) = 668.5$ N

When the lift accelerates when travelling up, applying Newton's second law, we have $ma = R - mg$.

Hence $R = ma + mg = (70 \times 0.4) + (70 \times 9.8) = 714$ N

When the lift is moving at constant speed, $R = 70g = 686$ N

Therefore, the greatest value of the reaction is 714 N.

Note that:
deceleration $= \frac{2}{8}$
$= 0.25 \text{ m s}^{-2}$

---

4 A lift of mass 1500 kg is ascending with an acceleration of 3 m s^{-2}.

(a) Calculate the tension in the lift cable.

(b) A lady with a mass of M kg stands on the floor of the lift. The magnitude of the reaction of the floor of the lift on the lady is 384 N. Calculate the mass of the lady to the nearest kg.

Answer

4 (a)

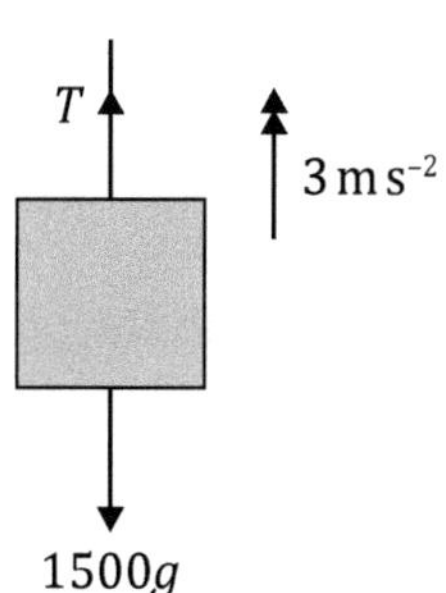

Applying Newton's 2nd law of motion, we have

$$ma = T - mg$$

$$1500 \times 3 = T - 1500 \times 9.8$$

Solving we obtain $T = 19\,200$ N

(b)

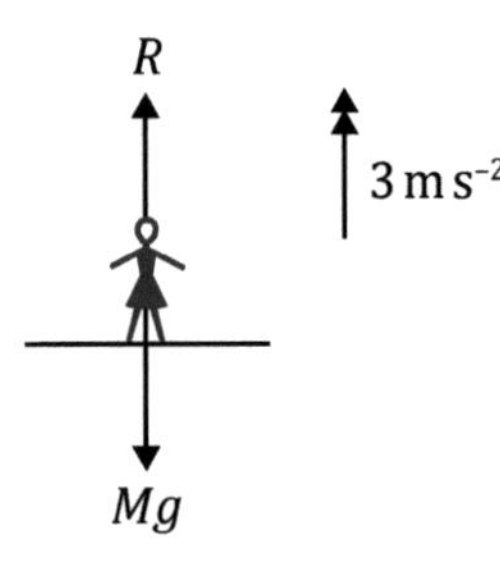

$$Ma = R - Mg$$
$$3M = 384 - 9.8M$$
$$12.8M = 384$$
$$M = 30 \text{ kg}$$

5 A lift is pulled upwards by means of a vertical cable. Initially, the lift is at rest. It then accelerates until it reaches a maximum speed. The lift moves at this maximum speed before decelerating uniformly at 3 m s^{-2} to rest. The total mass of the lift and its contents is 360 kg.

(a) Calculate the tension in the lift cable:

(i) when the lift is decelerating,

(ii) when the lift is moving at its maximum speed. [4]

(b) A crate on the floor of the lift has a mass of 25 kg. When the lift is accelerating the reaction between the crate and the floor of the lift is 280 N.

Find the magnitude of the acceleration of the lift. [3]

Answer

(a) (i)

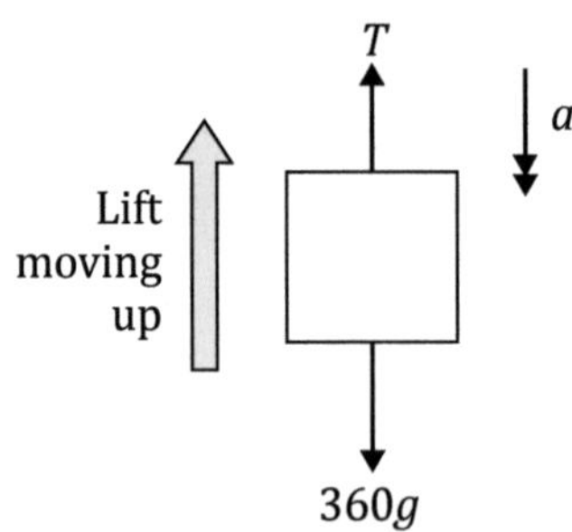

Applying Newton's second law in the vertical direction, we have

$$360a = 360g - T$$
$$360 \times 3 = (360 \times 9.8) - T$$

Giving $T = 2448$ N

(ii) When the lift is moving at constant speed there is no acceleration and therefore no resultant force.

Hence the tension is in equilibrium with the weight.

$$T = mg = 360 \times 9.8 = 3528 \text{ N}$$

(b)

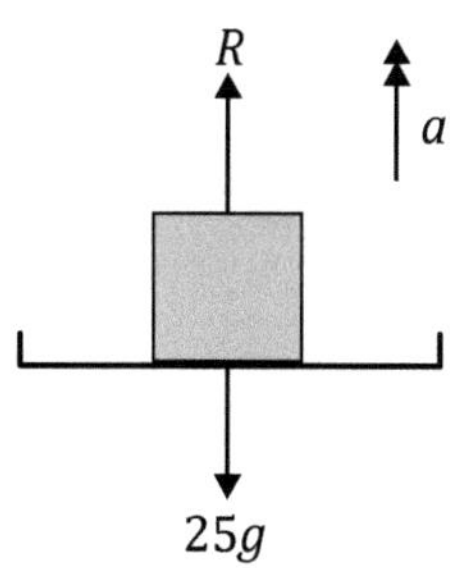

Applying Newton's second law in the vertical direction for the crate, we have

$$25a = R - 25g$$

$$25a = 280 - 25 \times 9.8$$

Hence $$a = 1.4 \text{ m s}^{-2}$$

8.4 The motion of particles connected by strings passing over pulleys or pegs

Pulleys and pegs

In this section you will be looking at string/rope passing over a smooth pulley or peg.

- A pulley is a wheel which is free to rotate. Normally we consider smooth pulleys that rotate without any friction.
- A smooth peg is a pin over which a string/rope passes and it can be assumed to behave exactly like a smooth pulley, so again we assume no frictional forces act.

Modelling assumptions

Certain modelling assumptions are made when answering questions. Make sure you use the ones that are applicable to the question being answered.

- The string connecting is assumed to be light so the tension is constant throughout its length.
- Pulleys and pegs are smooth so no frictional forces act.
- Horizontal surfaces on which masses rest are assumed to be smooth so there are no frictional forces acting.
- The string is inextensible which means it does not stretch which means both masses connected by the string will experience the same acceleration.
- There are no external forces acting which could alter tensions and accelerations.

When particles are connected by a string passing over a smooth pulley or peg, there is a constant tension in the string provided that the string is light. The forces acting on the particles due to the string are the tensions which act towards the pulley or the peg. The tensions acting on the pulley act in the opposite direction. Each particle can be treated separately and Newton's second law can be used to form an equation of motion connecting the acceleration to the forces acting.

When both particles are hanging freely

In the arrangement shown, two masses of 6 kg and 8 kg are connected by a light inextensible string passing over a smooth pulley. The masses are originally at rest and then released.

Accelerations are of equal magnitude because the string is inextensible.

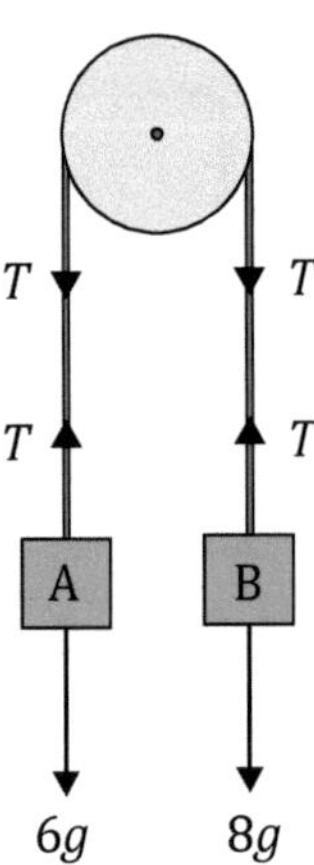

When the system shown above is released, the heavier particle (B in this case) will accelerate downwards and the other particle (A in this case) will accelerate upwards. Both accelerations will be equal in magnitude but opposite in direction.

Newton's second law of motion can be applied to each mass separately.

Applying Newton's second law of motion to mass A, we obtain

$$ma = T - 6g$$

so $$6a = T - 6g \qquad (1)$$

Applying Newton's second law of motion to mass B, we obtain

$$ma = 8g - T$$

so $$8a = 8g - \mathrm{T} \qquad (2)$$

You now have two equations, with two unknowns, which can be solved simultaneously to determine a and T.

Remember to check the two values satisfy equation (2) by substituting them in and checking that the left-hand side of the equation equals the right-hand side.

Adding equations (1) and (2) we have

$$14a = 2g$$

$$14a = 2 \times 9.8$$

Hence $$a = 1.4\ \mathrm{m\,s^{-2}}$$

Substituting this value for a into equation (1) we have

$$6 \times 1.4 = T - (6 \times 9.8)$$

Hence $$T = 67.2\ \mathrm{N}$$

When one particle is freely hanging and the other particle is on a smooth horizontal plane

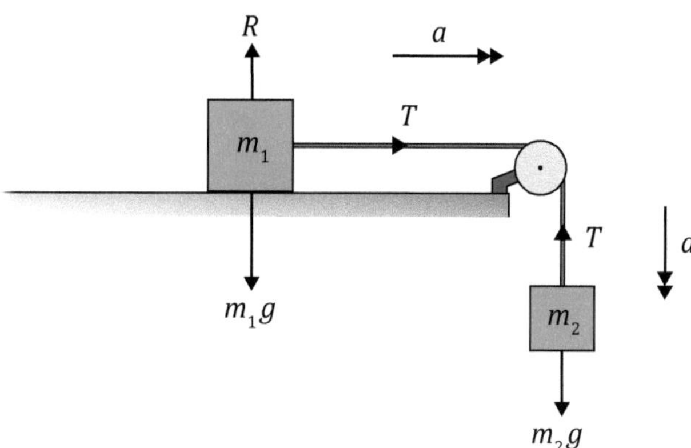

As there is no vertical motion for mass m_1 the vertical forces are in equilibrium.

Resolving vertically gives $R = m_1 g$

Applying Newton's second law of motion to mass m_1 we obtain

$$m_1 a = T \qquad (1)$$

Applying Newton's second law of motion to mass m_2 we obtain

$$m_2 a = m_2 g - T \qquad (2)$$

Then a and T are again found by solving the simultaneous equations (1) and (2).

Examples

1

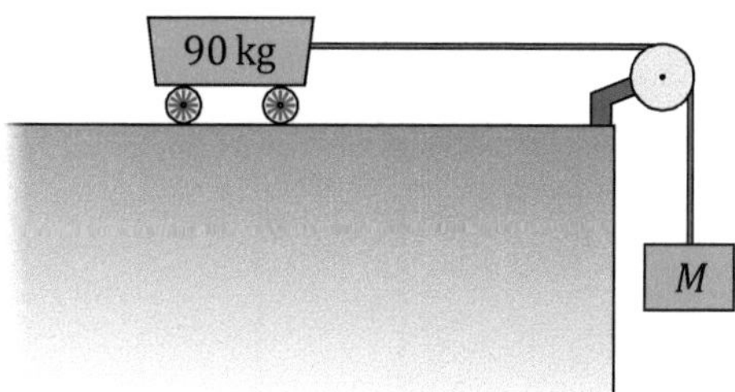

A truck of mass 90 kg is on a smooth horizontal surface. This truck is connected by a light inextensible string to a mass M kg which passes over a smooth pulley as shown in the above diagram. Mass M is free to move in a vertical direction.

The system is released from rest and the system accelerates with an acceleration of 4.8 m s^{-2}.

(a) Find the tension, T, in the string. [2]

(b) Find the value of mass M in kg. [3]

Answer

1 (a)

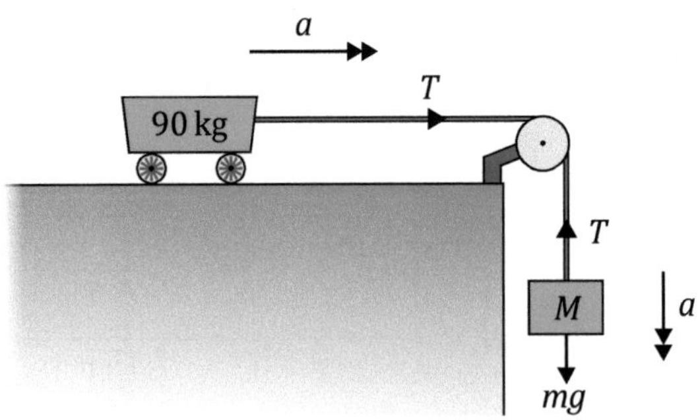

Applying Newton's 2nd law to the 90 kg mass, we have

$$T = 90 \times 4.8$$
$$= 432\text{ N}$$

(b) Applying Newton's 2nd law to mass M, we have

$$Ma = Mg - T$$

As $a = 4.8\text{ m s}^{-2}$

$$4.8M = 9.8M - T$$
$$T = 5M \qquad (1)$$

Substituting $T = 432$ N into equation (1) we have

$$432 = 5M$$

Hence $M = 86.4\text{ kg}$

---

2 Two identical buckets A and B are tied to a light inextensible rope which is passed over a smooth pulley as shown in the diagram.

Both buckets have a mass of 0.4 kg.

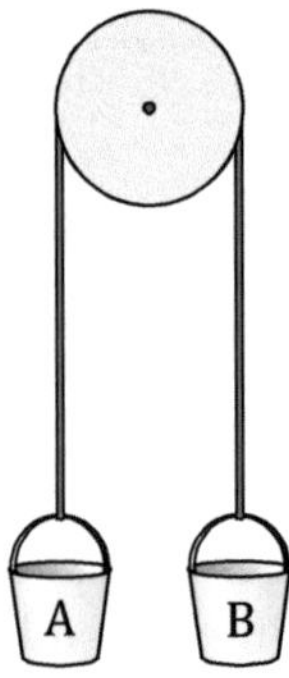

(a) State the magnitude of the tension in the rope. [2]

(b) Bucket A is held at rest while sand of mass 4.6 kg is added to bucket B. The system is then released from rest and bucket B travels downward with the sand inside.

(i) Using Newton's second law of motion, form two equations, one for each bucket. [4]

(ii) Solving the equations formed in part (i), find the acceleration of the system and the tension in the rope. [4]

Answer

(a)

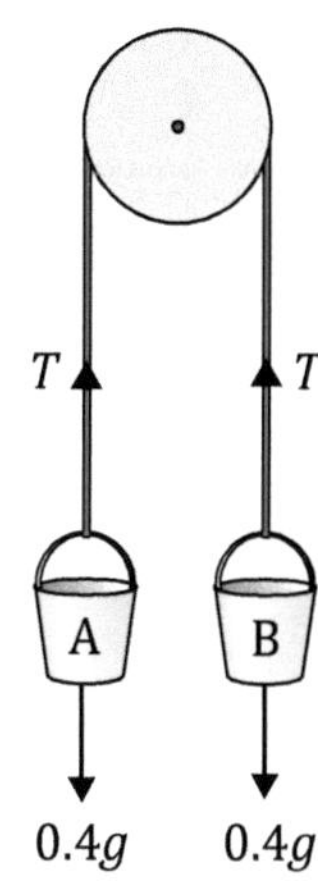

Resolving vertically for bucket A, we obtain

$$T = 0.4g = 0.4 \times 9.8 = 3.92 \text{ N}$$

(b)

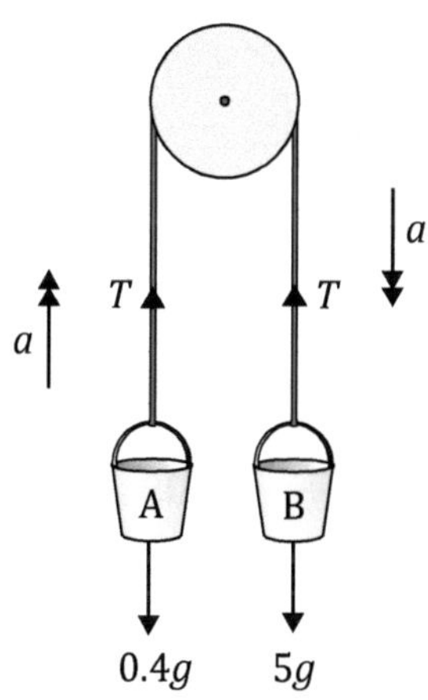

In this case, the combined mass of bucket B and sand is 5 g. Note that the tensions are different from those in part (a).

(i) Applying Newton's second law of motion to A, we obtain

$$ma = T - 0.4g$$

$$0.4a = T - 0.4g \qquad (1)$$

Note the different values of m for A and B.

Applying Newton's second law of motion to B, we obtain

$$ma = 5g - T$$

$$5a = 5g - T \qquad (2)$$

(ii) Solving equations (1) and (2) simultaneously, we obtain

$$a = 8.35 \text{ m s}^{-2} \text{ and } T = 7.26 \text{ N}$$

---

3 Two particles of mass 4 kg and 5 kg are connected by a light inextensible string passing over a smooth fixed pulley. The system is released from rest with the string taut and both masses 3 m above the ground. Find the velocity of the 4 kg mass when the 5 kg mass hits the ground. [5]

Answer

3

If you have values for the masses, you can always find the tension and the acceleration. Here the acceleration is needed so that the equations of motion can be used.

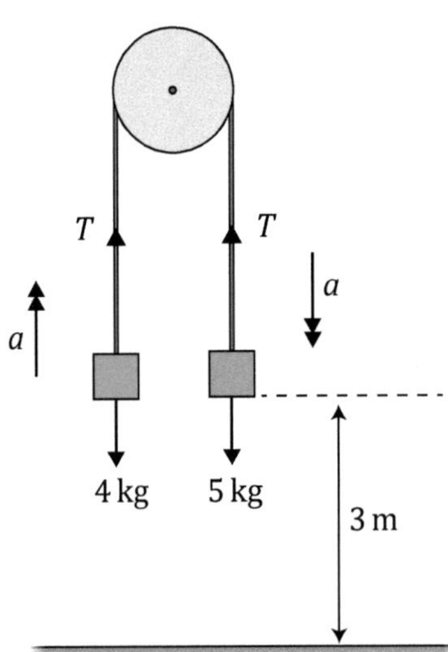

Applying Newton's 2nd law of motion to the 5 kg mass, we obtain

$$ma = mg - T$$

$$5a = 5g - T \qquad (1)$$

Applying Newton's 2nd law of motion to the 4 kg mass, we obtain

$$ma = T - mg$$

$$4a = T - 4g \qquad (2)$$

Solving equations (1) and (2) simultaneously for a we obtain

$$a = \frac{g}{9} = \frac{9.8}{9} = 1.08\ \text{m s}^{-2}$$

Using the equations of motion for the 4 kg mass which will accelerate upwards with an acceleration of $1.08\ \text{m s}^{-2}$, we have

$u = 0\ \text{m s}^{-1}$, $a = 1.08\ \text{m s}^{-2}$, $s = 3\text{m}$, $v = ?$

Using $v^2 = u^2 + 2as$

$$= 0^2 + 2 \times 1.08 \times 3$$

$$v = 2.6\ \text{m s}^{-1}$$

---

4 The diagram shows two objects A and B, of mass 5 g and 9 g respectively, connected by a *light* inextensible string passing over a smooth peg. Initially, the objects are held at rest. The system is then released.

(a) Find the magnitude of the acceleration of A and the tension in the string. [7]

(b) What assumption did the word 'light', emphasised in the first sentence, enable you to make in your solution? [1]

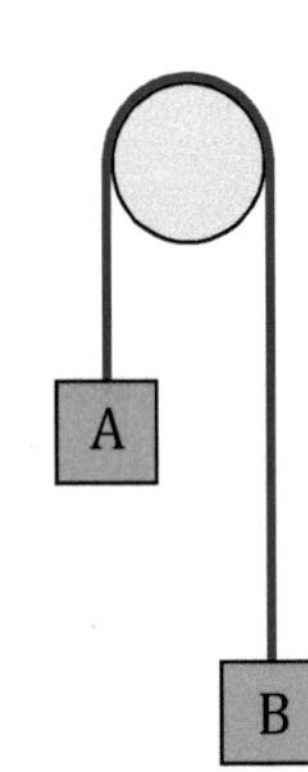

Answer

4 (a)

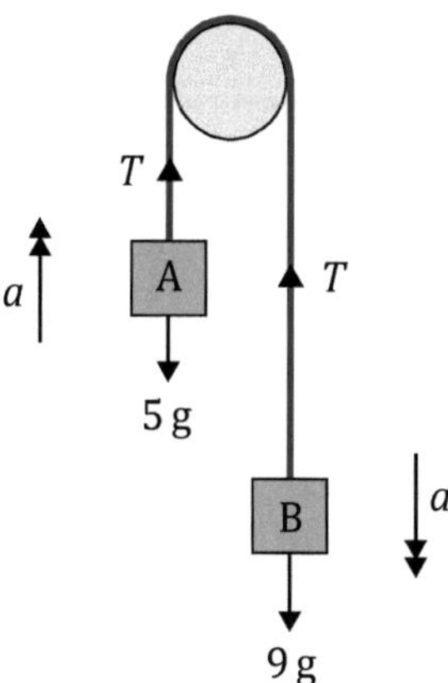

Applying Newton's second law of motion to A, we have

$$ma = T - 5g$$

$$5a = T - 5g \qquad (1)$$

Applying Newton's second law of motion to B, we have

$$ma = 9g - T$$

$$9a = 9g - T \qquad (2)$$

Adding equations (1) and (2), we obtain

$$14a = 4g$$

$$14a = 4 \times 9.8$$

Hence $\quad a = 2.8 \text{ m s}^{-2}$

From (1), $\quad T = 5a + 5g = 63 \text{ N}$

(b) The assumption of a light string enables us to regard the tension T as being constant throughout the string.

---

5 Two masses m kg and M kg where $M > m$, are connected by a light inextensible string passing over a smooth pulley. The system is released from rest with the two masses level with each other.

(a) Show that the acceleration of the M kg mass is given by

$$a = \frac{g(M - m)}{(M + m)} \qquad [4]$$

(b) If $M = 6$ kg and $m = 2$ kg, find the magnitude of the acceleration of the 2 kg mass. [2]

Answer

5 (a)

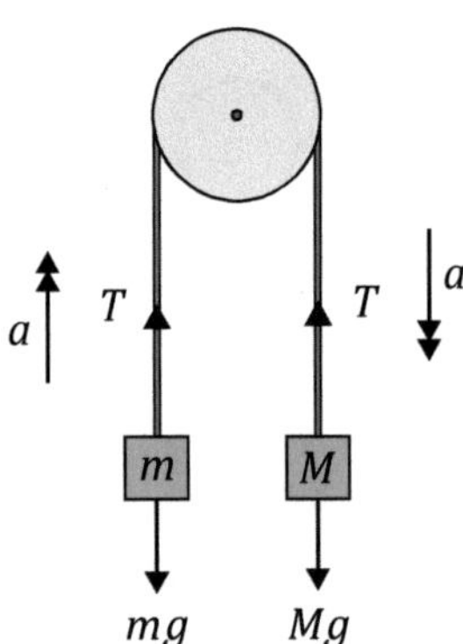

Applying Newton's 2nd law of motion to the M kg mass, we obtain

$$Ma = Mg - T \quad (1)$$

Applying Newton's 2nd law of motion to the m kg mass, we obtain

$$ma = T - mg \quad (2)$$

Adding equations (1) and (2) we obtain

$$Ma + ma = Mg - mg$$

$$a(M + m) = g(M - m)$$

$$a = \frac{g(M - m)}{(M + m)}$$

(b) When $M = 6$ kg and $m = 2$ kg the acceleration of the 6 kg mass is

$$a = \frac{9.8(6 - 2)}{(6 + 2)} = 4.9 \text{ m s}^{-2}$$

As the string is assumed to remain taut, both masses will have the same magnitude of acceleration.

Hence the acceleration of the 2 kg mass = 4.9 m s^{-2}

Step by STEP

A truck of mass 180 kg runs on smooth horizontal rails. A light inextensible rope is attached to the front of the truck. The rope runs parallel to the rails until it passes over a light smooth pulley. The rest of the rope hangs down a vertical shaft. When the truck is required to move, a load of M kg is attached to the end of the rope in the shaft and the brakes are then released.

(a) Find the tension in the rope when the truck and the load move with an acceleration of magnitude 0.8 m s^{-2} and calculate the corresponding value of M. [5]

(b) In addition to the assumptions given in the question, write down one further assumption that you have made in your solution to this problem and explain how that assumption affects both of your answers. [3]

Steps to take

1 Draw a diagram and mark on the forces acting. As the rails and pulley are smooth, there are no frictional forces. Mark on the diagram, the direction of the acceleration.

2 As the rope is light and inextensible this means that the tension in the rope will remain constant.

3 There are two connected particles. Newton's 2nd law of motion can be applied to each particle in turn to obtain two equations.

4 The two equations can be solved simultaneously to find the tension and the acceleration.

5 Think about the assumptions not given in the question (e.g. absence of frictional forces due to air resistance, etc.).

Answer

(a)

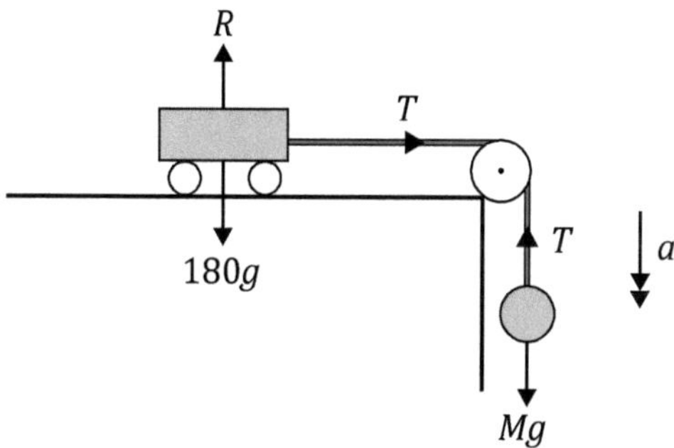

Applying Newton's 2nd law to the truck

$$180 \times 0.8 = T$$
$$T = 144\text{ N}$$

Applying Newton's 2nd law to the load

$$Ma = Mg - T$$
$$T = M(g - a)$$
$$144 = M(9.8 - 0.8)$$
$$M = 16\text{ kg}$$

(b) There is no resistance to the motion due to external forces (e.g. air resistance).

If there was air resistance on the load it would act in the same direction as the tension (i.e. in the opposite direction to the motion).

The resistance will oppose the motion meaning the downward force would increase which means M would have to increase to keep the acceleration at 0.8 m s^{-2}.

If M is kept the same, the tension would be lower thus keeping the upward forces at their original value.

Could also have the answer that the truck and load are both modelled as particles.

Test yourself

1. A particle of mass 3 kg is attached to the end of a light inextensible string and allowed to hang vertically. The string and particle are accelerated vertically downwards with an acceleration of 2 m s^{-2}.

   Calculate the tension in the string.

2. Two particles of mass 1.5 kg and 2 kg are connected by a light inextensible string that passes over a smooth peg. The particles are released from rest as shown in the diagram.

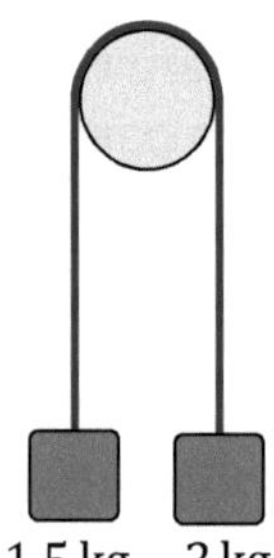

   (a) By using Newton's second law of motion, form an equation of motion for each particle and hence show that the tension in the string is 16.8 N.

   (b) Find the acceleration of the system.

   (c) If the particles are initially at the same level, determine the speed of the heavier particle when the distance between the two particles is 1 m.

3. Particles A and B are connected by a light inextensible string which moves over a smooth pulley and this system is shown in the following diagram.

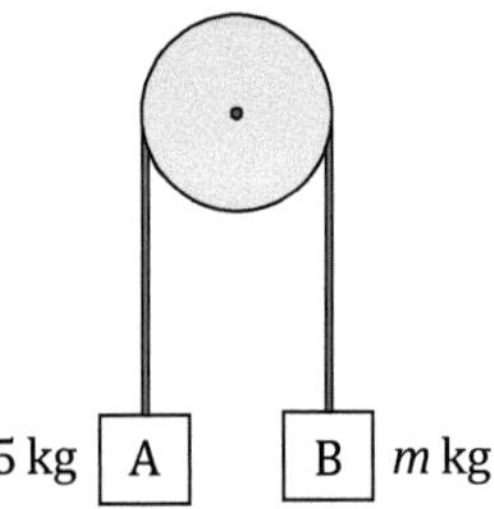

   The mass of particle A is 5 kg and the mass of particle B is m kg where $m > 5$.

   The particles are released from rest with the strings taut and vertical.

   When the particles are released, the magnitude of the common acceleration of the particles is 2 m s^{-2}.

   (a) Find the value of the tension in the string.

   (b) Find the value of m.

4. The diagram below shows two masses P and Q connected by a light inextensible string passing over a smooth pulley. Mass P is held at rest with the string just taut and it is then released.

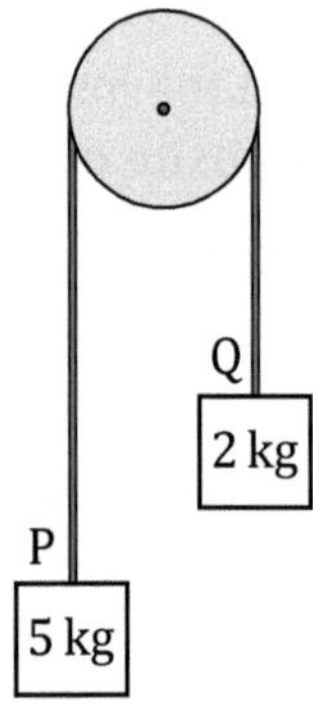

(a) Calculate the magnitude of the acceleration of P and the tension in the string.
(b) What assumption does the word light in the description of the string enable you to make in your solution?

5 Two masses, one of 3 kg and the other of 1 kg are connected by a light inextensible string passing over a smooth pulley. Both strings are taut and the 3 kg mass is released from rest at a height of 2 m above the ground.

(a) Find the velocity with which the 3 kg mass hits the ground.
(b) Explain what modelling assumptions you have made in your answer to part (a).

6 A lift starts from rest and ascends with a uniform acceleration of 4 m s^{-2} until it reaches a velocity of 10 m s^{-1}. It travels at this constant velocity for 3 s and then is brought to rest by a constant deceleration of 2.5 m s^{-2}. A person with a mass of 55 kg is standing on the floor of the lift during its motion.

(a) Sketch a velocity–time graph of the lift's motion.
(b) Calculate the total distance travelled by the lift.
(c) Determine the maximum magnitude of the reaction of the floor of the lift on the person during the lift's motion.

7 A person of mass 45 kg stands on the floor of a lift of mass 455 kg which is ascending with an acceleration of a m s^{-2}. The tension in the lift cable is 6000 N.

(a) Find the magnitude of the acceleration a.
(b) Find the reaction between the floor of the lift and the man.

8 A box of mass 5 kg rests on a smooth horizontal table. This mass is connected by a light inextensible string which passes over a smooth peg at the end of the table, to a 2 kg mass which hangs freely.
The 2 kg mass is released from rest with the string taut.
Find the acceleration of the two masses and the tension in the string.

9 The diagram below shows two bodies P and Q, of masses 8 kg and 5 kg respectively. The two bodies are connected by a light inextensible string passing over a smooth pulley. The larger mass, P, lies on the table and the lighter mass, Q, hangs freely below the pulley.

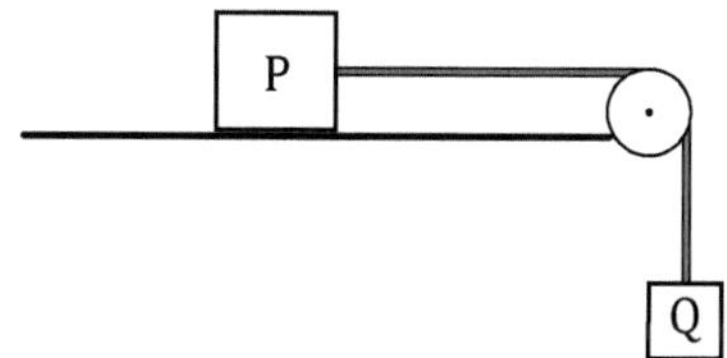

Initially the system is at rest with the string taut. The system is then released.

(a) Find the acceleration of the two masses and the tension in the string.
(b) An additional mass M kg is added to mass Q and the system is again released from rest. This causes the acceleration of the two masses to increase to 5 m s^{-2}. Find the size of mass M.

Summary

Newton's second law of motion

Unbalanced forces produce an acceleration according to the equation

force = mass × acceleration or for short $F = ma$

Lifts accelerating, decelerating and travelling with constant velocity

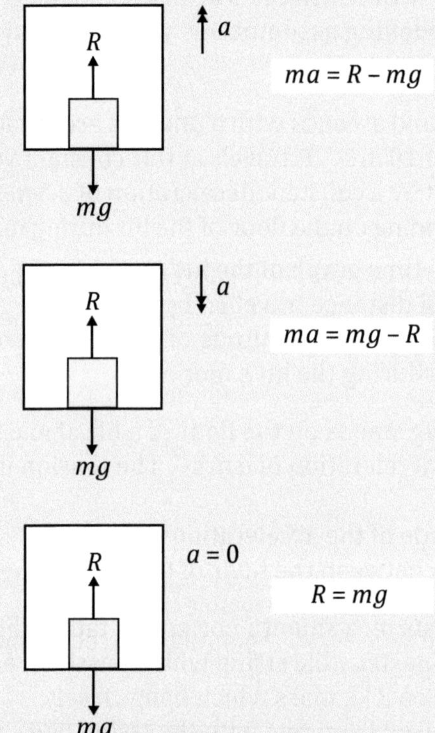

The motion of particles connected by strings passing over fixed pulleys or pegs

Pulleys or pegs are smooth so no frictional forces act.

Strings are light and inextensible so the tension remains constant.

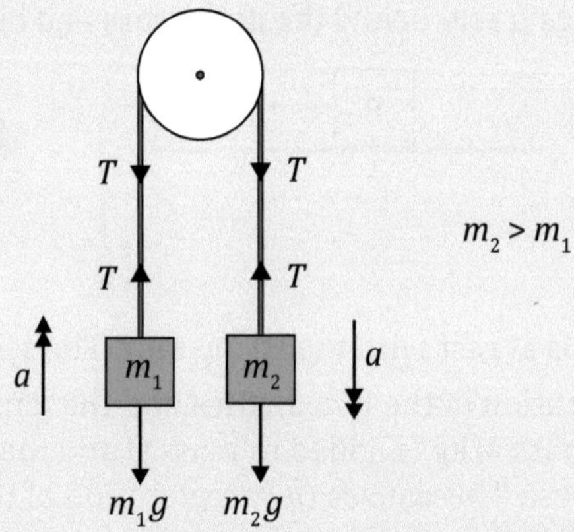

Acceleration of each mass is the same as the string is taut.

Newton's second law can be applied to each mass separately.

For m_1, $m_1 a = T - m_1 g$

while for m_2, $m_2 a = m_2 g - T$

9 Vectors

Introduction

Vectors were covered in Pure AS Topic 9 so you should quickly look over this topic as a refresher.

Vectors are important in mechanics as many of the systems in mechanics act in two or three dimensions. In this topic we will only be dealing with systems in two dimensions.

This topic covers the following:

9.1 Calculating the magnitude and direction of a vector and converting between component form and magnitude/direction form

9.2 Using vectors to solve problems

9.1 Calculating the magnitude and direction of a vector and converting between component form and magnitude/direction form

Vectors and scalars

Scalar quantities have magnitude only whereas vector quantities have both magnitude and direction.

Here is a table showing which quantities are scalar and which are vectors.

Scalar quantity	Vector quantity
Distance	Displacement
Speed	Velocity
	Force
	Acceleration

It is important to note that distance is the magnitude of displacement and speed is the magnitude of velocity.

You will notice that in the book and on the examination paper you will see vector quantities written in bold like this:

These are the commonest letters used for these quantities but any other letter can be used.

F for a force

v for a velocity

r or **s** for a displacement

a for an acceleration

Notice that the letters used for vectors are in bold and also not in italics.

When handwriting vectors in an exam, instead of using bold you can underline the letter to show it is a vector, so $\underline{a}$, $\underline{E}$, $\underline{v}$ and $\underline{x}$ are all vectors.

Forces given as 2D-vectors

Forces are vector quantities and can be expressed in terms of the perpendicular unit vectors **i** and **j**.

For example, a force vector can be written as $\mathbf{F} = 12\mathbf{i} + 5\mathbf{j}$

Calculating the magnitude of a vector

Any vector can be represented as a straight line and the magnitude of the vector is the length of the line.

The magnitude of a vector **a** can be written as the modulus of **a** which is $|\mathbf{a}|$.

Suppose we have the following vector for velocity

$$\mathbf{v} = a\mathbf{i} + b\mathbf{j}$$

The magnitude of vector **v** is given by $|\mathbf{v}| = \sqrt{a^2 + b^2}$

Example

1 A particle is travelling with a velocity given by $\mathbf{v} = 12\mathbf{i} + 5\mathbf{j}$ m s^{-1}. Calculate the magnitude of the velocity.

Answer

1 $|\mathbf{v}| = \sqrt{a^2 + b^2} = \sqrt{12^2 + 5^2} = \sqrt{144 + 25} = \sqrt{169} = 13$ m s^{-1}

Finding the direction of a vector

A vector has both magnitude and direction. Usually the direction is the angle made by the vector and the unit vector **i** but be guided by the direction indicated in the question.

Suppose we want to find the direction made by the vector $\mathbf{F} = 2\mathbf{i} + 5\mathbf{j}$ and the unit vector **i** giving the answer to one decimal place.

First draw a diagram

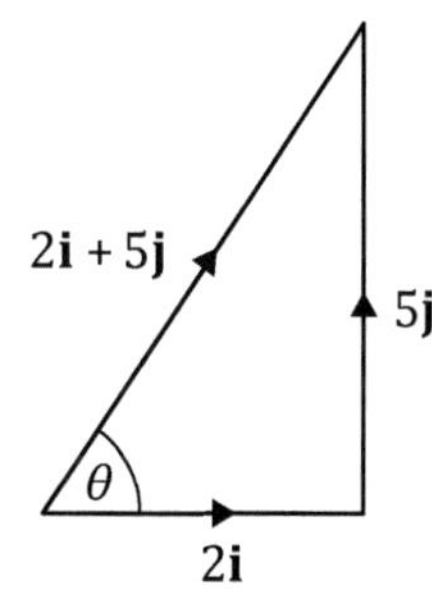

The unit vectors **i** and **j** are at right angles to each other so the triangle is a right-angled triangle.

Now use trigonometry to work out angle θ.

$$\tan\theta = \frac{5}{2}$$

$$\theta = \tan^{-1}\left(\frac{5}{2}\right)$$

$$= 68.2° \text{ (1 d.p.)}$$

$\tan\theta = \dfrac{\text{opposite}}{\text{adjacent}}$

Finding the resultant of two or more forces in vector form

Forces in two dimensions can be expressed using the unit vectors **i** and **j** or as column vectors.

So two vectors could be expressed as

$$\mathbf{F} = 2\mathbf{i} + 3\mathbf{j} \text{ and } \mathbf{G} = 3\mathbf{i} - 4\mathbf{j}$$

or as

$$\mathbf{F} = \begin{pmatrix} 2 \\ 3 \end{pmatrix} \text{ and } \mathbf{G} = \begin{pmatrix} 3 \\ -4 \end{pmatrix}$$

To find the resultant of two or more forces in vector form you simply add the vectors together.

Resultant of **F** and **G** = $(2\mathbf{i} + 3\mathbf{j}) + (3\mathbf{i} - 4\mathbf{j})$

$$= 5\mathbf{i} - \mathbf{j} \quad \text{or} \quad \begin{pmatrix} 2 \\ 3 \end{pmatrix} + \begin{pmatrix} 3 \\ -4 \end{pmatrix} = \begin{pmatrix} 5 \\ -1 \end{pmatrix}$$

Examples

1 Three forces **F**, **G** and **H** act at point P.

If **F** = **i** + 10**j**, **G** = 2**i** + 8**j** and **H** = 4**i** + 7**j** find the resultant force at P.

Answer

1 Resultant force = **F** + **G** + **H** = (**i** + 10**j**) + (2**i** + 8**j**) + (4**i** + 7**j**) = 7**i** + 25**j**

---

2 A robot sets off from O and moves with the following two consecutive displacements; −4**i** + 7**j** and 6**i** − 5**j**. Find the final displacement of the robot from O.

Answer

2 Total displacement = (−4**i** + 7**j**) + (6**i** − 5**j**) = 2**i** + 2**j**

---

3 The following three forces all act at the same point:

−5**i** + 2**j**
−7**i** − 6**j**
9**i** + 4**j**

The unit vectors **i** and **j** act horizontally and vertically respectively.

Find the fourth force, **F**, that would need to be added to the three forces in order to make the resultant force zero.

Answer

3 Resultant force = (−5**i** + 2**j**) + (−7**i** − 6**j**) + (9**i** + 4**j**) = −3**i**

So **F** = 3**i**

To make 0, 3**i** would need to be added to the resultant force (i.e. −3**i** + 3**i** = 0).

---

4 Two forces **P** and **Q** act on an object such that

P = 2**i** + 3**j**
Q = 3**i** − **j**

The object has a mass of 10 kg. Calculate the magnitude and direction of the acceleration of the object giving your answers to 1 decimal place.

The unit vector **i** is directed along the *x*-axis and the unit vector **j** is directed along the *y*-axis.

Answer

4 We need to find the resultant force vector which we shall call **R**. This is obtained by adding vectors **P** and **Q**.

R = **P** + **Q** = (2**i** + 3**j**) + (3**i** − **j**) = 5**i** + 2**j**

Use Pythagoras's theorem to work out the magnitude (i.e. size) of the resultant.

$$\mathbf{R}^2 = 5^2 + 2^2$$
$$= 25 + 4$$
$$= 29$$
$$\mathbf{R} = \sqrt{29}\text{ N}$$
$$= 5.385\text{ N} = 5.4\text{ N (1 d.p.)}$$

BOOST Grade

Ensure you give the answer to the accuracy specified in the question and make sure you include the correct units.

The resultant force can be shown on a diagram and the angle is marked between the resultant force and the unit vector **i**.

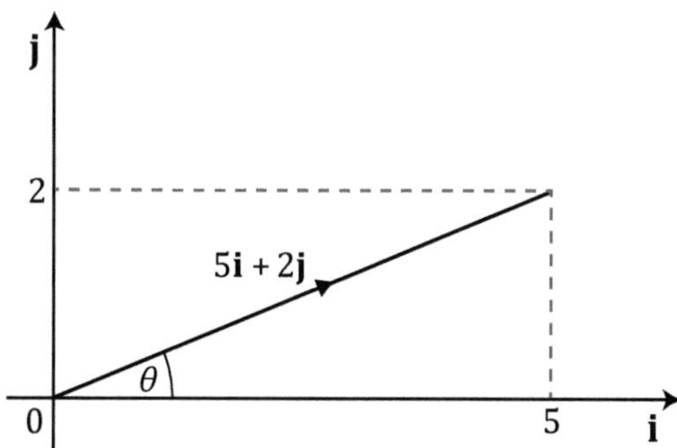

Now find angle θ by using the angle the resultant makes with the positive x-axis (i.e. the direction to the unit **i** vector).

$$\theta = \tan^{-1}\left(\frac{2}{5}\right)$$
$$= 21.8° \text{ (1 d.p.) to the positive } x\text{-axis.}$$

Work out the acceleration using the formula $F = ma$

$F = ma$ is Newton's 2nd law of motion.

$$a = \frac{F}{m} = \frac{5.385}{10} = 0.5 \text{ ms}^{-2} \text{ (1 d.p.)}$$

The direction of the acceleration is the same as that of the force.

Hence, direction is

21.8° (1 d.p.) to the positive x-axis (or direction of the unit vector **i**).

Finding the resultant of several vectors expressed as column vectors

Instead of the vectors $3\mathbf{i} - \mathbf{j}$ and $\mathbf{i} + 7\mathbf{j}$

we can write them as the column vectors $\begin{pmatrix} 3 \\ -1 \end{pmatrix}$ and $\begin{pmatrix} 1 \\ 7 \end{pmatrix}$

To add two or more column vectors to find the resultant, you simply add the top numbers and add the bottom numbers like this:

$$\begin{pmatrix} 3 \\ -1 \end{pmatrix} + \begin{pmatrix} 1 \\ 7 \end{pmatrix} = \begin{pmatrix} 3+1 \\ -1+7 \end{pmatrix} = \begin{pmatrix} 4 \\ 6 \end{pmatrix}$$

Examples

1 Point P starts from the origin O and is then subject to the following three consecutive displacements $\begin{pmatrix} 0 \\ -4 \end{pmatrix}$, $\begin{pmatrix} 5 \\ -1 \end{pmatrix}$ and $\begin{pmatrix} -3 \\ 9 \end{pmatrix}$.

(a) Find the final displacement of P from O.

(b) Find the displacement of O from P.

Answer

1 (a) Final displacement $\overrightarrow{OP} = \begin{pmatrix} 0 \\ -4 \end{pmatrix} + \begin{pmatrix} 5 \\ -1 \end{pmatrix} + \begin{pmatrix} -3 \\ 9 \end{pmatrix} = \begin{pmatrix} 2 \\ 4 \end{pmatrix}$

The final displacement is the sum of the individual displacements.

(b) Displacement $\overrightarrow{PO} = -\overrightarrow{OP} = \begin{pmatrix} -2 \\ -4 \end{pmatrix}$

2 A point P is subject to the following consecutive displacements.

$$7\mathbf{i} - 2\mathbf{j} \text{ and } 8\mathbf{i} + 10\mathbf{j}$$

Find the total displacement and the total distance moved giving your answer to the nearest metre.

Answer

2 The two consecutive displacements can be shown on the following set of axes:

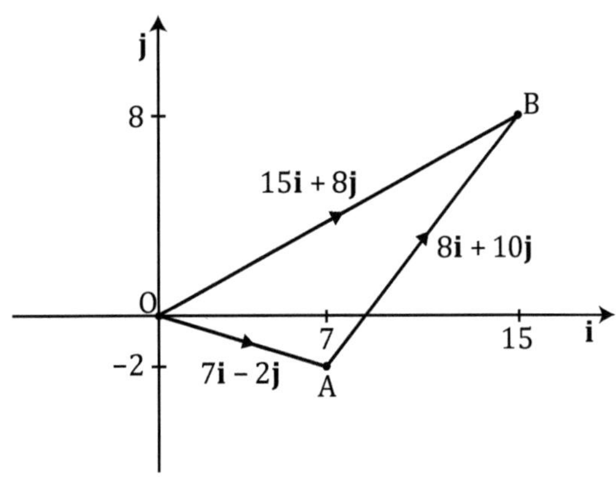

Notice that the individual displacements are added to give the total displacement.

$$\text{Total displacement} = (7\mathbf{i} - 2\mathbf{j}) + (8\mathbf{i} + 10\mathbf{j}) = 15\mathbf{i} + 8\mathbf{j}$$

To find the total distance it is necessary to find the separate distances OA and AB and then add these together. If you wanted the straight line distance OB you could use the total displacement vector and find its magnitude.

$$\begin{aligned}\text{Total distance moved} &= \sqrt{7^2 + (-2)^2} + \sqrt{8^2 + 10^2} \\ &= \sqrt{53} + \sqrt{164} \\ &= 20 \text{ m (nearest m)}\end{aligned}$$

Do not use the total displacement vector to work out the distance moved as this will give the straight line distance from the starting point (i.e. the distance OB on the graph).

3 A boat is searching for a wreck starting from point A and then travelling to points B and C before finding the wreck at D.

The displacement from A to B is $3\mathbf{i} + 4\mathbf{j}$ km

The displacement from B to C is $5\mathbf{i} + 4\mathbf{j}$ km

The displacement from C to D is $12\mathbf{i} + 5\mathbf{j}$ km

(a) Find the magnitude of the displacement from A to D giving your answer to one decimal place.

(b) Find the distance travelled by the boat from A to D giving your answer to one decimal place and explain why this is not equal to the magnitude of the displacement from A to D.

Answer

Column vectors have been used here but you could have used **i** and **j** vectors instead.

3 (a) $\overrightarrow{AD} = \overrightarrow{AB} + \overrightarrow{BC} + \overrightarrow{CD}$

$$\overrightarrow{AD} = \begin{pmatrix}3\\4\end{pmatrix} + \begin{pmatrix}5\\4\end{pmatrix} + \begin{pmatrix}12\\5\end{pmatrix} = \begin{pmatrix}20\\13\end{pmatrix}$$

$$|\overrightarrow{AD}| = \sqrt{20^2 + 13^2} = 23.9 \text{ km (1 d.p.)}$$

(b) Total distance from A to D = $|\overrightarrow{AB}| + |\overrightarrow{BC}| + |\overrightarrow{CD}|$

$$|\overrightarrow{AB}| = \sqrt{3^2 + 4^2} = 5.0\text{ km}$$

$$|\overrightarrow{BC}| = \sqrt{5^2 + 4^2} = 6.4\text{ km}$$

$$|\overrightarrow{CD}| = \sqrt{12^2 + 5^2} = 13.0\text{ km}$$

Total distance = 5.0 + 6.4 + 13.0 = 24.4 km (1 d.p.)

The magnitude of the displacement from A to D is measured along the straight line joining points A and D. The distance is the distances for the individual sections (i.e. A to B then B to C and then C to D) which are further than the direct straight line route from A to D.

9.2 Using vectors to solve problems

Vectors can be used to solve problems in two dimensions. When dealing with vectors you need to think carefully about the problem and not get confused between a vector and the magnitude of a vector. If no diagram is given in the question, feel free to draw one yourself as it will often help to understand the arrangement being described.

Step by STEP

Two forces **F** and **G** acting on an object are such that

$$\mathbf{F} = \mathbf{i} - 8\mathbf{j},$$

$$\mathbf{G} = 3\mathbf{i} + 11\mathbf{j}.$$

The object has a mass of 3 kg. Calculate the magnitude and direction of the acceleration of the object. [7]

Steps to take

1 Find the resultant force by adding the two vectors together.

2 Use Newton's 2nd law of motion, **F** = *m***a** to find the vector for the acceleration.

3 Find the magnitude of the acceleration.

4 Use trigonometry and the acceleration vector to determine the required angle.

Answer

Resultant force = **F** + **G**

$$= \mathbf{i} - 8\mathbf{j} + 3\mathbf{i} + 11\mathbf{j}$$

$$= 4\mathbf{i} + 3\mathbf{j}$$

Using Newton's 2nd law: $\mathbf{F} = m\mathbf{a}$

$$4\mathbf{i} + 3\mathbf{j} = 3\mathbf{a}$$

$$\mathbf{a} = \frac{4}{3}\mathbf{i} + \mathbf{j}$$

$$|\mathbf{a}| = \sqrt{\left(\frac{4}{3}\right)^2 + 1^2}$$

$$= \frac{5}{3}\text{ m s}^{-2}$$

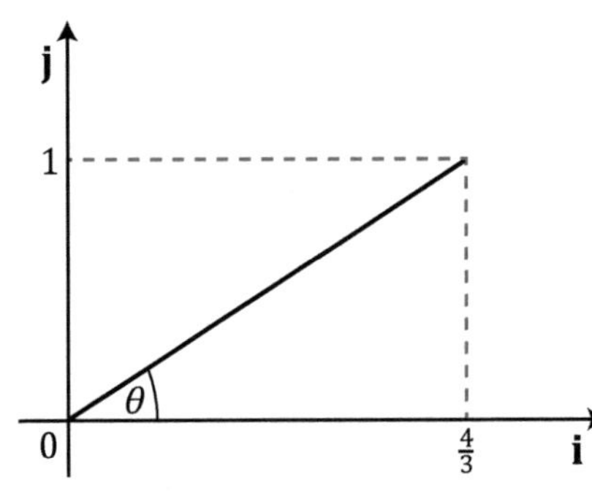

$$\tan\theta = \frac{1}{\left(\frac{4}{3}\right)}$$

$$\tan\theta = \frac{3}{4}$$

$$\theta = 36.87° \text{ (to the } \mathbf{i} \text{ vector)}$$

Examples

1 Three forces **F**, **G** and **H** act at a point where:

$$\mathbf{F} = 3\mathbf{i} + 5\mathbf{j}$$

$$\mathbf{G} = -\mathbf{i} + \mathbf{j}$$

$$\mathbf{H} = a\mathbf{i} + b\mathbf{j}$$

If the resultant of these three forces is zero:

(a) Find the values of a and b.

(b) Write down the vector for force **H**.

Answer

The sum of the vectors in the **i** direction is equal to zero as is the sum of the vectors in the **j** direction.

1 (a) $\mathbf{F} + \mathbf{G} + \mathbf{H} = (3\mathbf{i} + 5\mathbf{j}) + (-\mathbf{i} + \mathbf{j}) + (a\mathbf{i} + b\mathbf{j})$

Now $\mathbf{F} + \mathbf{G} + \mathbf{H} = 0$

So $(2\mathbf{i} + a\mathbf{i}) + (6\mathbf{j} + b\mathbf{j}) = 0$

Hence $a = -2$ and $b = -6$

(b) $\mathbf{H} = -2\mathbf{i} - 6\mathbf{j}$

---

2 The velocity of a particle is given by $5\mathbf{i} + 12\mathbf{j}$.

Find:

(a) The speed of the particle.

(b) The angle made by the direction of the particle to the unit vector **i** giving your answer to the nearest degree.

Answer

Remember speed is a scalar quantity and is the magnitude of the velocity.

(a) $\mathbf{v} = 5\mathbf{i} + 12\mathbf{j}$

Speed $= |\mathbf{v}| = \sqrt{5^2 + 12^2} = \sqrt{25 + 144} = 13 \text{ m s}^{-1}$

(b)

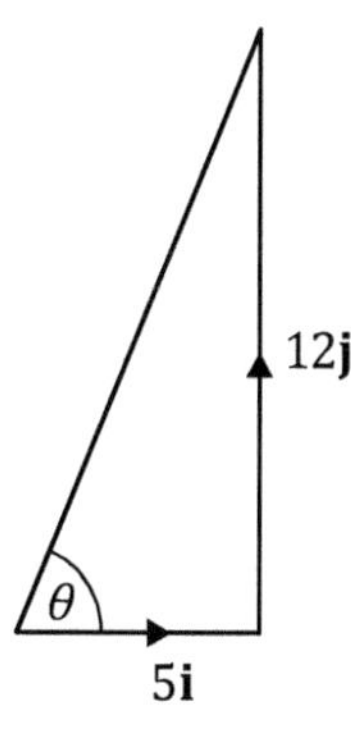

$$\tan\theta = \frac{12}{5} = 2.4$$

$$\theta = \tan^{-1} 2.4 = 67° \text{ (nearest degree)}$$

Step by STEP

A drone sets off on a journey from point A to point B travelling at constant height in a straight line where $\overrightarrow{AB} = 2\mathbf{i} + \mathbf{j}$ km. It then continues at the same height and in a straight line to point C where $\overrightarrow{BC} = 4\mathbf{i} - 3\mathbf{j}$ km.

(a) Write down the displacement vector for A to C.

(b) Write down the vector which represents the straight line path from C to A.

(c) Find the total distance travelled by the drone on its journey from A to C.

Steps to take

1 Add the vector for the journey from A to B to the vector for the journey from B to C and simplify.

2 This is the reverse of the vector from A to C so simply reverse the signs.

3 Need to find the distance from A to B and then the distance from B to C using the magnitude of a vector formula. These two distances are added together.

Answer

(a) $\overrightarrow{AC} = \overrightarrow{AB} + \overrightarrow{BC} = (2\mathbf{i} + \mathbf{j}) + (4\mathbf{i} - 3\mathbf{j}) = 6\mathbf{i} - 2\mathbf{j}$

(b) $\overrightarrow{CA} = -\overrightarrow{AC} = -(6\mathbf{i} - 2\mathbf{j}) = -6\mathbf{i} + 2\mathbf{j}$

(c) $|\overrightarrow{AB}| = \sqrt{2^2 + 1^2} = \sqrt{5} = 2.24$ km

$|\overrightarrow{BC}| = \sqrt{4^2 + (-3)^2} = \sqrt{25} = 5$ km

Total distance travelled = 2.24 + 5 = 7.24 km (2 d.p.)

Finding a vector from its magnitude and direction

If the magnitude and direction are given but not the actual vector, then the vector can be found using the method shown here.

BOOST Grade

Watch out for the words 'exact value' in questions. Any roots of numbers which when worked out would result in the need to approximate would no longer give exact values. Instead do not work the roots out.

Exact values for the angles must be used. This was covered in the *AS Pure* book. Look back if you can't remember how to work the angles as exact values.

Notice the question asks for exact values for a and b. This means that you must keep the square root and not work it out as a decimal.

Example

1 A force of 12 N acts at a point O at an angle of 30° to the unit **i** vector. Express the force as a vector in the form $a\mathbf{i} + b\mathbf{j}$ where a and b are **exact** values to be found.

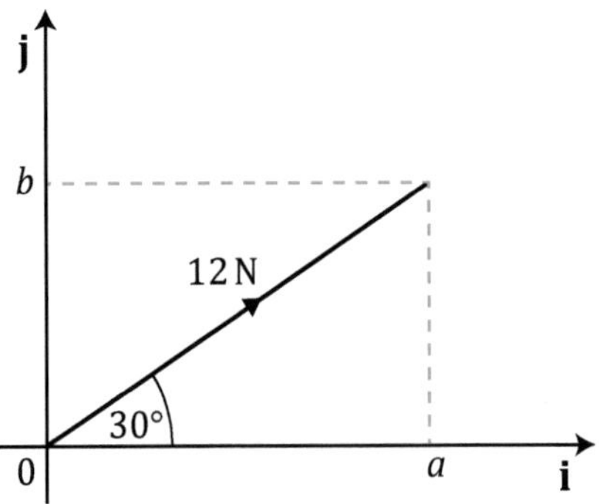

Answer

1 Using trigonometry:

$$\frac{a}{12} = \cos 30°$$

$$a = 12 \cos 30° = 12 \times \frac{\sqrt{3}}{2} = 6\sqrt{3}$$

$$\frac{b}{12} = \sin 30°$$

$$b = 12 \sin 30° = 12 \times \frac{1}{2} = 6$$

Vector is $6\sqrt{3}\mathbf{i} + 6\mathbf{j}$

Test yourself

1. Two forces **L** and **M** are acting on an object.
   $\mathbf{L} = 5\mathbf{i} + 9\mathbf{j}$
   $\mathbf{M} = 2\mathbf{i} + 15\mathbf{j}$
   If the mass of the object is 5 kg, find:
   (a) The acceleration **a**.
   (b) The magnitude of the acceleration.
   (c) The direction of the acceleration to the **i** vector, to the nearest degree.

2. The three forces, $2\mathbf{i} - 3\mathbf{j}$, $\mathbf{i} + 6\mathbf{j}$ and $-4\mathbf{i} - 4\mathbf{j}$, act on a particle together with a fourth force, **F**.
   If the resultant force is $7\mathbf{i} + 2\mathbf{j}$, find **F**.

3. Starting from O a particle moves in such a way that it follows the following consecutive displacement vectors:
   $3\mathbf{i} - 4\mathbf{j}$, $2\mathbf{i} + \mathbf{j}$, $-4\mathbf{i} + 6\mathbf{j}$, $5\mathbf{i} - \mathbf{j}$ and $5\mathbf{i} + 3\mathbf{j}$
   What is the final displacement of the particle from O?

4. Find in vector form, the acceleration **a** produced by a 10 kg mass which is subjected to the forces $(4\mathbf{i} + \mathbf{j})$ N and $(6\mathbf{i} - 6\mathbf{j})$ N.

5. If **i** is the unit vector due east and **j** is the unit vector due north and the forces $(3\mathbf{i} + 6\mathbf{j})$ N, $(-2\mathbf{i} - \mathbf{j})$ N and $(2\mathbf{i} - \mathbf{j})$ N act on a particle situated at the origin, find:
   (a) The resultant force acting on the particle in the form $a\mathbf{i} + b\mathbf{j}$.
   (b) The magnitude of the resultant force.
   (c) The bearing of the resultant force to the nearest degree.

6. A particle is subjected to the following three forces:
   $(\mathbf{i} + 3\mathbf{j})$, $(2\mathbf{i} - 5\mathbf{j})$ and $(4\mathbf{i} + 6\mathbf{j})$, where all the forces are measured in newtons.
   (a) Find the vector for the resultant force.
   (b) If the mass of the particle is 2 kg, find the vector for the acceleration **a**.
   (c) Find the magnitude of the acceleration to the nearest whole number and the angle it makes to the unit vector **i** to the nearest 0.1°.

7. A boat sails in straight lines from X to Y and then from Y to Z.
   The displacement from X to Y is given by $-2\mathbf{i} + 13\mathbf{j}$ km.
   The displacement from Y to Z is given by $7\mathbf{i} - \mathbf{j}$ km.
   (a) Find the displacement from X to Z.
   (b) Find the magnitude of the displacement from X to Z.
   (c) Find the total distance the boat has sailed from X to Z giving your answer to 3 significant figures.

8. A particle P leaves the origin O with a speed of 15 m s^{-1} and travels in a straight line with this constant speed at an angle of 60° to the positive x-direction.
   Find the vector, **v**, for the velocity of the particle expressing it in the form $\mathbf{v} = a\mathbf{i} + b\mathbf{j}$ where a and b are **exact** values to be found.

Summary

Scalar quantities have magnitude (i.e. size) only and include distance and speed.

Vector quantities have both magnitude and direction and include displacement, velocity, acceleration and force.

Vectors are typed in bold and not in italics, so **s**, **r**, **v**, **a** and **F** are all vectors.

The resultant of vectors acting at a point can be found by adding the individual vectors.

The magnitude of a vector

The vector $\mathbf{r} = a\mathbf{i} + b\mathbf{j}$ has magnitude given by $|\mathbf{r}| = \sqrt{a^2 + b^2}$

The direction of a vector

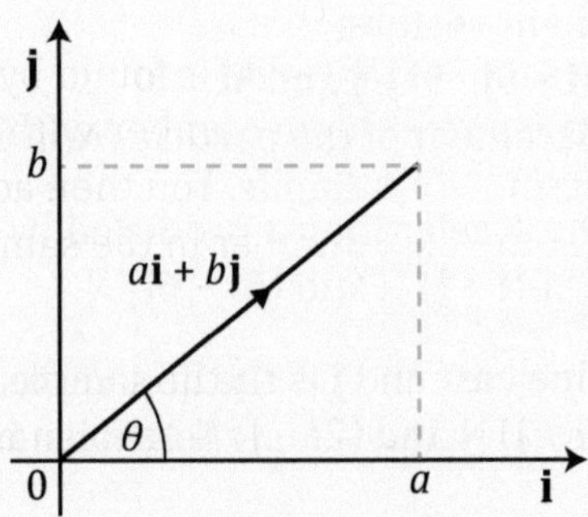

The angle made by the vector $a\mathbf{i} + b\mathbf{j}$ to the unit vector $\mathbf{i}$ is θ, where $\theta = \tan^{-1}\left(\frac{b}{a}\right)$

Converting from magnitude and direction to a vector

If you know the magnitude and direction of a vector you can convert this to vector form in the following way

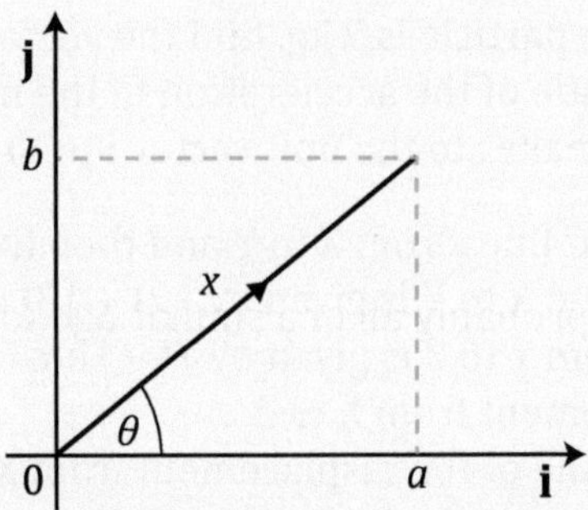

If the length of the vector is x and it is inclined at an angle θ to the **i**-direction (or positive x-axis) then using trigonometry

$$\frac{a}{x} = \cos\theta°, \text{ so } a = x\cos\theta° \quad \text{and} \quad \frac{b}{x} = \sin\theta°, \text{ so } b = x\sin\theta°$$

Vector is $\mathbf{x} = a\mathbf{i} + b\mathbf{j}$

Test yourself answers

Section A Statistics

Topic 1

1 (a) There are several problems with the sample. Any of the following or similar answers:
Her sample is too small – it is taken over a single week. The newspaper's sample could have been taken every day in June over several years.
The sample should have been taken evenly spread out over the month of June and not just the first week.
(b) Any two from the following:
Take data spread out over the whole of June and not just the first week.
Collect data every day for the entire month of June.
Use historical data obtained from weather records over many years.

2 (a) sampling interval $= \frac{\text{population}}{\text{sample size}} = \frac{300}{30} = 10$
A random number is used between 1 and 10 by picking a number out of a hat or generating a random number using a calculator/program/website. Suppose the random number was 5. The fifth person entering the club is asked to complete the questionnaire. Then every 10th person entering the club is asked to complete the questionnaire.
(b) If a random number such as 298 was picked initially you would need to wait until they arrived for the first questionnaire or find them when the club was full.

3 (a) Opportunity sampling (i.e. she has used the easiest sample – her friends).
(b) The sample may be biased because:
Her friends are probably all of a similar age to herself and won't reflect the population.
Her friends are likely to have things in common with her which may include watching similar soaps.

4 Random sampling

5 (a) The population is the whole group so in this case it would be all the households in the road.
A sample is a selection of households in the road.
(b) A sample would be quicker to perform.
A sample would be less expensive to take.
(c) Use systematic sampling using the house numbers from 1 to 350 or a simple random sample of houses.
If systematic sampling was chosen:

$$\text{Sampling interval} = \frac{\text{population}}{\text{sample size}} = \frac{350}{50} = 7$$

If simple random sampling was chosen as the method you would pick random numbers in the range 1 to 350 using random number tables/a calculator/a website. If the number has already been picked you would pick again. This is repeated until the 50 random house numbers have been obtained.

Obtain a random number using calculator, website, etc., from 1 to 7 and use this household as the start for the sample.
Now keep adding 7 to this number to get the next householder in the sample until 50 households are in the sample.

6 (a) The population would be all the supporters attending the match.
(b) It would be impractical/impossible to get all the supporters to take part in the survey.
It would be too expensive in terms of people to do the survey, cost of forms, etc.
It would take lots of time to process all the results.
It would take too long for the surveys.
(c) (i) Opportunity sampling
(ii) The sample could consist of a group of away supporters who arrived at the same time (e.g. on a coach) and thus be unrepresentative of the population.
There could also be a bias towards men or women if a group came together.

Topic 2

1 (a) Strong positive correlation. Students who did well in chemistry also did well in physics.
(b) Interpolation is using the graph inside the range of the data to find a missing value. For example, we can use the scatter diagram to find the likely mark in physics for a student who got a mark of 70% in chemistry but was absent for the physics exam.
(c) As the maximum chemistry mark was 90% you would be using the regression equation beyond what you had values for (i.e. extrapolation). It might be possible to get a physics mark greater than 100% using the equation, which is an impossible mark.

2 (a) (i) LQ = 22.5 kg
(ii) UQ = 25.4 kg
(iii) Median = 24.1 kg
(b) IQR = 25.4 − 22.5 = 2.9 kg
It represents the spread of the middle half of the data (i.e. one quarter either side of the median).

3 (a) This is the length when the mass on the spring is zero, so it is the original length of the spring.
(b) Length (cm) = 20 + 0.025 × 100 = 22.5 cm.
(c) The regression equation should not be used too far from the range of masses used for the experiment. 5 kg far exceeds the maximum mass used and there is no guarantee that the regression equation holds for this mass.

4 (a) True as there will be a corresponding temperature for each value of latitude. Each pair of values is a data plot so there will be 30 in total.
(b) True. If one thing causes another and vice versa, they will be correlated.
(c) False. The gradient is positive so a high value of one variable corresponds to a high value of the other variable.
(d) True. The two variables show negative correlation It is reasonable to

assume that if there are lots of firework-related accidents, it is because there were not many organised displays so people were more likely to buy their own fireworks..

(e) False. The line of best fit will have a positive gradient but an individual point can be either side of the trend line so it would be possible to be tall and have a lower arm length. Remember that the trend looks at all the points and not just one.

(f) True. As the general trend means as one variable goes up the other goes down.

(g) True. The different outcomes show the drug affected the patient so there is causation.

5 $\frac{\sum x_i}{n} = \frac{102}{20} = 5.1$

Variance $= \frac{\sum x_i^2}{n} - \left(\frac{\sum x_i}{n}\right)^2 = \frac{580}{20} - 5.1^2 = 2.99$

Standard deviation $= \sqrt{\text{variance}} = \sqrt{2.99} = 1.73$ (3 s.f.)

6 (a) Mean, $\mu = \frac{\sum x_i}{n} = \frac{92}{20} = 4.60$

Standard deviation, $\sigma = \sqrt{\frac{\sum x_i^2}{n} - \left(\frac{\sum x_i}{n}\right)^2}$

$= \sqrt{\frac{435.42}{20} - \left(\frac{92}{20}\right)^2}$

$= 0.782$

Summary statistics

Masses of a certain species of bird (kg)

Mass of bird	N	Mean	Standard Deviation	Mini-mum	Lower quartile	Median	Upper quartile	Maxi-mum
	20	4.60	0.782	3.8	4.05	4.40	5.08	6.8

(b) IQR $= Q_3 - Q_1 = 5.08 - 4.05 = 1.03$

An outlier is a value larger than $Q_3 + 1.5 \times \text{IQR} = 5.08 + 1.5 \times 1.03 = 6.625$

As mass 6.8 kg > 6.625 kg this value is an outlier.

(c) An outlier is a value smaller than $Q_1 - 1.5 \times \text{IQR}$

$Q_1 - 1.5 \times \text{IQR} = 4.05 - 1.5 \times 1.03 = 2.505$

A mass of 2.505 kg is the smallest mass without it being an outlier.

7 (a) (i) Positive correlation

(ii) The greater the number of grams of sugar in a cup of cereal, the greater the number of calories.

(b) (i) The gradient is the number of grams of sugar per calorie.

(ii) The graph should really only be used for values within the range 100 – 230 calories. 350 would involve extrapolating the graph and the graph may not behave in the same way.

(iii) It might be causal but not all calories come from sugar as some come from fat and protein. For example, some cereals contain nuts which are high in fat and hence high in calories.

Topic 3

The probability of event A only is found by subtracting the probability of $P(A \cap B)$ from $P(A)$. The probability of B only is also found by subtracting $P(A \cap B)$ from $P(B)$.

1 (a) Probability of event A only $= 0.45 - 0.25 = 0.2$
Probability of event B only $= 0.30 - 0.25 = 0.05$

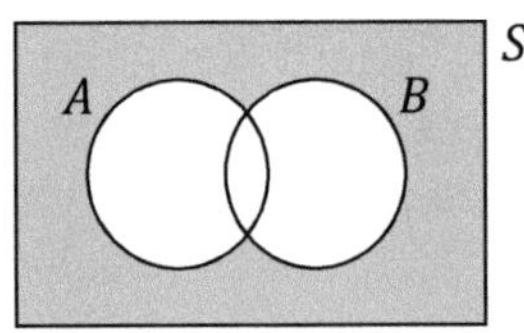

A Venn diagram showing these probabilities and the probability of $A \cap B$ can now be drawn.

$P(A \cup B) = 0.2 + 0.25 + 0.05 = 0.5$

Alternatively, you could use

$$P(A \cup B) = P(A) + P(B) - P(A \cap B)$$
$$= 0.45 + 0.30 - 0.25$$
$$= 0.5$$

$A' \cap B'$ represents everything outside $A \cup B$. Note that the probability of both of these events adds up to 1. Hence, $P(A' \cap B') = 1 - P(A \cup B)$.

(b) $P(A' \cap B') = 1 - 0.5 = 0.5$

2 (a) (i) When throwing two dice, the sample space consists of 36 pairs of scores.
The number of pairs the same = 6
Hence, P(scores are equal) = 6/36 = 1/6

All the possible scores where Amy's score is higher than Bethany's score are listed. Here it is assumed that Amy throws first so the pairs are listed in the order (A, B).

(ii) The possible scores where Amy's score is higher are:
(2, 1)
(3, 1), (3, 2)
(4, 1), (4, 2), (4, 3)
(5, 1), (5, 2), (5, 3), (5, 4)
(6, 1), (6, 2), (6, 3), (6, 4), (6, 5)

Hence, Probability that Amy's score is higher $= \frac{15}{36} = \frac{5}{12}$

Note that the sample space is now reduced from 36 to only 3 of which there is only one pair the same giving a total of 4 (i.e. (2, 2)).

(b) Sample space consists of:
(1, 3), (3, 1), (2, 2)
Probability scores are equal $= \frac{1}{3}$

Note that $P(A \cap B) = P(A) \times P(B)$ as they are independent.

3 (a) Using $P(A \cup B) = P(A) + P(B) - P(A \cap B)$ and $P(A \cap B) = P(A) \times P(B)$, we obtain

$$0.5 = 0.3 + P(B) - 0.3 \times P(B)$$
$$0.2 = 0.7\,P(B)$$

Hence, $P(B) = 0.2857$ (correct to 4 d.p.)

Note that the probability of B only is given by $P(B) - P(A \cap B)$.

(b) P(exactly one event occurs) $= P(A \cup B) - P(A \cap B)$
$= 0.5 - 0.3 \times 0.2857$
$= 0.5 - 0.0857$
$= 0.4143$

4 (a)

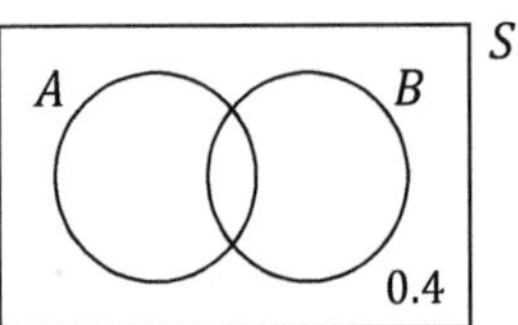

Note that in drawing this Venn diagram we have assumed that there is an overlap between A and B. If the events are mutually exclusive then there would be no overlap.

From the Venn diagram, $P(A \cup B) = 1 - 0.4 = 0.6$
Using $P(A \cup B) = P(A) + P(B) - P(A \cap B)$,
we obtain $0.6 = 0.4 + 0.35 - P(A \cap B)$.

Hence, $P(A \cap B) = 0.15$ and as there is an overlap between events A and B, the two events are not mutually exclusive.

(b) $P(A) \times P(B) = 0.4 \times 0.35 = 0.14$, and as $P(A \cap B) = 0.15$, we have $P(A) \times P(B) \neq P(A \cap B)$ proving that events A and B are not independent.

5 (a) As the events are independent we can use $P(A \cap B) = P(A) \times P(B)$.

$$P(A \cap B) = P(A) \times P(B)$$
$$= 0.4 \times 0.3 = 0.12$$

(b) $P(A \cup B) = P(A) + P(B) - P(A \cap B)$
$= 0.4 + 0.3 - 0.12 = 0.58$

(c) The probability that neither A nor B occur is given by $1 - P(A \cup B)$.
Probability that neither A nor B occur $= 1 - 0.58$
$= 0.42$

6 (a) Using $P(A \cup B) = P(A) + P(B) - P(A \cap B)$, we obtain

$$0.5 = 0.4 + 0.2 - P(A \cap B)$$
$$P(A \cap B) = 0.1$$

(b) $P(A) \times P(B) = 0.4 \times 0.2 = 0.08$
As $P(A \cap B) \neq P(A) \times P(B)$ the events A and B are not independent.

7 (a) Using $P(A \cup B) = P(A) + P(B) - P(A \cap B)$, we obtain

$$P(A \cap B) = P(A) + P(B) - P(A \cup B)$$
$$= 0.2 + 0.4 - 0.52$$
$$= 0.08$$

If the events A and B are independent, then probability of both events occurring

$$= P(A) \times P(B) = 0.2 \times 0.4 = 0.08$$

As $P(A) \times P(B) = P(A \cap B)$, events A and B are independent.

(b) $P(A \text{ only}) = P(A) - P(A \cap B)$
$= 0.2 - 0.08$
$= 0.12$

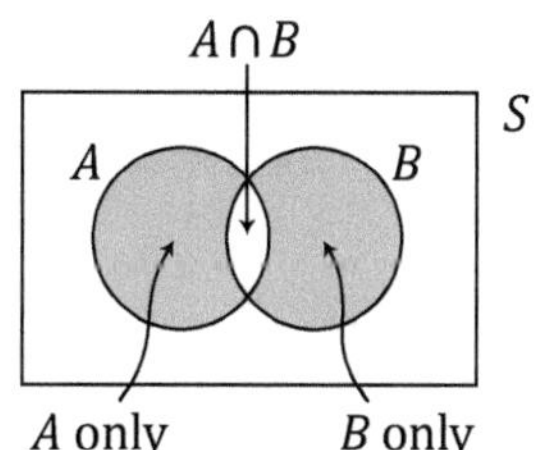

$P(B \text{ only}) = P(B) - P(A \cap B)$
$= 0.4 - 0.08$
$= 0.32$

$P(A \text{ or } B \text{ only}) = P(A \text{ only}) + P(B \text{ only}) = 0.12 + 0.32 = 0.44$

8 (a) Events A or B can occur but not both. There is no overlap.

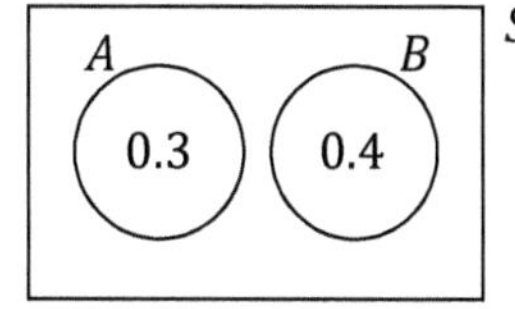

$P(A \cup B) = P(A) + P(B)$
$= 0.3 + 0.4 = 0.7$

(b)

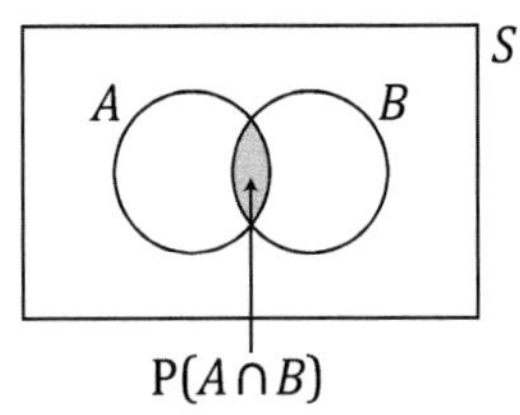

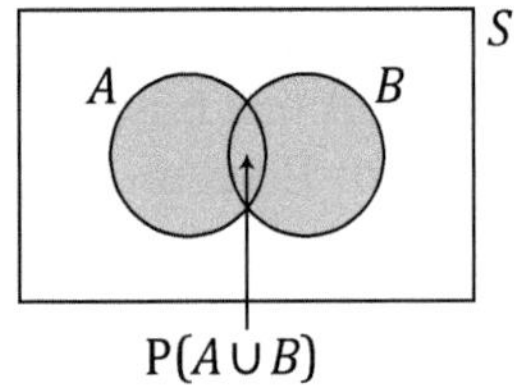

$P(A \cap B) = P(A) \times P(B) = 0.3 \times 0.4 = 0.12$
$P(A \cup B) = P(A) + P(B) - P(A \cap B)$
$= 0.3 + 0.4 - 0.12$
$= 0.58$

Mutually exclusive means both events cannot happen so there is no intersection between P(*A*) and P(*B*).

9 (a) $P(A \cup B) = P(A) + P(B)$
$= 0.2 + 0.3$
$= 0.5$

$P(A \cap B) = P(A) \times P(B)$

(b) $P(A \cup B) = P(A) + P(B) - P(A \cap B)$
$= 0.2 + 0.3 - 0.2 \times 0.3$
$= 0.2 + 0.3 - 0.06$
$= 0.44$

$A \subset B$ means that set *A* is a subset of set *B* so the union will just be the set of *B*.

(c) $P(A \cup B) = P(B) = 0.3$

Topic 4

The formula is obtained from the formula booklet.

1 (a) $P(X = x) = \binom{n}{x}p^x(1 - p)^{n-x}$
$p = 0.25$ and $n = 20$.
$P(X = 4) = \binom{20}{4}0.25^4(1 - 0.25)^{20-4}$
$= \binom{20}{4}0.25^4(0.75)^{16}$
$= 0.1897$ (correct to 4 s.f.)

By tables:
$P(X = 4) = P(X \le 4) - P(X \le 3)$
$= 0.4148 - 0.2252$
$= 0.1896$

(b) The binomial distribution tables are used here with $n = 20$, $p = 0.25$ and $x = 7$.
P(fewer than 8 bulbs) = $P(X \le 7) = 0.8982$

Notice the way $e^{-3.4}$ can be taken out as a factor to simplify the calculation.

2 (a) (i) $P(X = 4) = e^{-3.4}\frac{3.4^4}{4!} = 0.1858$ (correct to 4 s.f.)

(ii) $P(X \le 2) = e^{-3.4}\left(1 + 3.4 + \frac{3.4^2}{2}\right) = 0.3397$ (correct to 4 s.f.)

The Poisson cumulative distribution function tables are used to look up the $P(X \le x)$ values.

(b) $P(4 \le X \le 7) = P(X \le 7) - P(X \le 3)$
$= 0.9769 - 0.5584$
$= 0.4185$ (correct to 4 s.f.)

3 Mean $\lambda = np = 100 \times 0.08 = 8$
X is Po(8)
$P(X < 5) = P(X \le 4) = 0.0996$

The Poisson distribution function table is used here to find $P(X \le 4)$ with $x = 4$, and mean $\lambda = 8$.

4 (a) (i) $P(X = x) = \binom{n}{x} p^x (1-p)^{n-x}$
$p = 0.4$, $n = 20$ and $x = 10$.
$P(X = 10) = \binom{20}{10} 0.4^{10}(1 - 0.4)^{20-10}$
$P(X = 10) = \binom{20}{10} 0.4^{10}(0.6)^{10}$
$= 0.1171$ (correct to 4 s.f.)

The binomial distribution B(20, 0.4) is used here. The probability $p = 0.4$ is too high for the Poisson distribution to be used where ideally p should be less than 0.1. Also, n should be > 50.

(ii) $P(X > 7) = 1 - P(X \le 7)$
$= 1 - 0.4159$
$= 0.5841$

To find $P(X \le 7)$ we use the tables for the binomial distribution function with $n = 20$, $p = 0.4$ and $x = 7$.

(b) $\lambda = np = 300 \times 0.05 = 15$
X is distributed as Po(15)
$P(X < 10) = P(X \le 9) = 0.0699$

Tables should be used here as using the formula would be tedious because you would have 10 individual probabilities to calculate before adding them together.

5 (a) (i) Let X be the number of defective cups
$p = \frac{5}{100} = 0.05$ and $n = 50$
$P(X = x) = \binom{n}{x} p^x (1-p)^{n-x}$
$P(X = 2) = \binom{50}{2} 0.05^2(1 - 0.05)^{50-2}$
$= \binom{50}{2} 0.05^2(0.95)^{48}$
$= 0.2611$

This formula is obtained from the formula booklet.

(ii) $P(3 \le X \le 8) = P(X \le 8) - P(X \le 2)$
Using binomial distribution tables with $n = 50$ and $p = 0.05$
$P(X \le 8) = 0.9992$ and $P(X \le 2) = 0.5405$
$P(2 \le X < 8) = 0.9992 - 0.5405 = 0.4587$

(b) (i) $p = \frac{1.5}{100} = 0.015$, which is less than 0.1 and $n = 250$ which is greater than 50, so the Poisson distribution can be used.

(ii) Let Y be the number of defective plates.
$p = \frac{1.5}{100} = 0.015$, $n = 250$ and $np = 250 \times 0.015 = 3.75$
$P(Y = x) = e^{-\lambda}\frac{\lambda^x}{x!}$
$P(Y = 4) = e^{-3.75}\frac{3.75^4}{4!}$
$= 0.194$

6 (a) (i) $P(X = x) = \binom{n}{x} p^x (1-p)^{n-x}$
$P(X = 6) = \binom{20}{6} 0.2^6(1 - 0.2)^{20-6}$
$= \binom{20}{6}(0.2)^6(0.8)^{14} = 0.109$

Note that $P(X \leq 8)$ finds the probability of X being 0, 1, 2, 3, 4, 5, 6, 7 and 8 and $P(X \leq 1)$ finds the probability of being 0 or 1. So by subtracting, we obtain the required probability.

(ii) $P(2 \leq X \leq 8) = P(X \leq 8) - P(X \leq 1)$
Using binomial distribution tables with $n = 20$ and $p = 0.2$,
$P(X \leq 8) = 0.9900$ and $P(X \leq 1) = 0.0692$
So $P(2 \leq X \leq 8) = 0.9900 - 0.0692$
$= 0.921$

(b) $\lambda = np = 200 \times 0.0123 = 2.46$

$$P(Y = 3) = e^{-2.46}\frac{2.46^3}{3!}$$
$$= 0.212$$

BOOST Grade

Do not use the tables for a single value. If you had used tables you would not have got any marks for part (b) of this question.

7 (a) (i) Let the number of seeds producing red flowers = X
X is distributed as B(20, 0.7)

$$P(X = 15) = \binom{20}{15}(0.7)^{15}(1 - 0.7)^{20-15}$$
$$= \binom{20}{15}(0.7)^{15}(0.3)^5$$
$$= 0.179$$

The binomial distribution function tables are used to find this value.

(ii) Let the number of seeds not producing red flowers = X'
X' is distributed as B(20, 0.3)
Now $P(X > 12) = P(X' < 8)$
$= 0.772$

The binomial distribution function tables cannot be used here as the maximum value of n is 50 and here it is 150. Poisson is used as an approximation.

(b) Let the number of seeds producing white flowers = Y
Y is distributed as B(150, 0.09)
The Poisson approximation is used here with mean
$\lambda = np = 150 \times 0.09 = 13.5$
$B(150, 0.09) \sim Po(13.5)$

The Poisson formula is looked up in the formula booklet.

$$P(Y = 10) = e^{-13.5} \times \frac{13.5^{10}}{10!}$$
$$= 0.076$$

Topic 5

1 (a) The test statistic here is X, the number of heads obtained.
(b) (i) $H_0 : p = 0{\cdot}5$
(ii) $H_1 : p > 0{\cdot}5$
(c) **Answer using p-values**
Need to find the number X of heads for the probability to exceed 0.05 of obtaining that number of heads.
Using B(9, 0.5)
$$P(X \leq x) = 1 - 0.05 = 0.95$$
From the tables $P(X \leq 6) = 0.9102$ and $P(X \leq 7) = 0.9805$
$P(X \geq 8) = 1 - P(X \leq 7) = 1 - 0.9805 = 0.0195$ which is less than 0.05
$P(X \geq 7) = 1 - P(X \leq 6) = 1 - 0.9102 = 0.0898$ which is greater than 0.05
From these results, it means that if 8 or more heads are tossed then at the 5% level of significance the coin is biased towards heads so the null hypothesis should be rejected.

(c) **Answer using critical values**
Using B(9, 0.5)
As there is a '>' sign in the alternative hypothesis, we need to consider the upper tail of the probability distribution.
We need to look at the tables and find the first probability that exceeds 0.95, then the critical value is the value after it.
From the table $P(X \leq 7) = 0.9805$ so the critical value is 8 and the critical region is $X \geq 8$.
There would have to be 8 or more heads for the null hypothesis to be rejected.

2 The null hypothesis is that the spinner is not biased towards 5 so the probability of obtaining a 5 is equal to $\frac{1}{5}$ or 0.2.

$$\mathbf{H_0}: p = 0.2$$

The alternative hypothesis is that the spinner is biased towards 5.

$$\mathbf{H_1}: p > 0.2$$

The test statistic is X, the number of fives obtained in 10 spins.
Distribution is B(10, 0.2)

Method making use of a critical region
As the alternative hypothesis is $\mathbf{H_1}: p > 0.2$ we are using the upper tail of the distribution.
Using B(10, 0.2) we find the first probability in the column that exceeds 0.95 and its corresponding X value and the critical value a is the first value after that.
From the table $P(X \leq 4) = 0.9672$ so the critical value a is 5 and the critical region is $X \geq 5$.
As $X = 4$ does not lie in the critical region, there is insufficient evidence at the 5% level of significance to suggest that the spinner is biased.

Alternative method making use of *p*-values
$P(X \geq 4) = 1 - P(X \leq 3) = 1 - 0.8791 = 0.1209$
The significance level = 5% = 0.05.
Now 0.1209 is greater than 0.05, so there is not enough evidence to reject the null hypothesis that the spinner is biased towards 5. So it is concluded that the spinner is unlikely to be biased towards 5.

3 **Method using critical value**
$\mathbf{H_0}: p = 0.4 \quad \mathbf{H_1}: p < 0.4$
X is B(20, 0.4)
As there is a '<' sign in the alternative hypothesis, we are using the lower tail of the probability distribution.
Using the tables to find the first value for the probability greater than 0.05 we obtain $P(X \leq 4) = 0.0510$ and the critical value is the value one more before this, so a, the critical value is 3 and the critical region is $X \leq 3$.
As 4 is not in the critical region there is not enough evidence to reject the null hypothesis. Hence the manager's statement is unlikely to be correct.

Method using *p*-values
$\mathbf{H_0}: p = 0.4 \quad \mathbf{H_1}: p < 0.4$
Test statistic X is the number of callers waiting more than 5 minutes.
X is B(20, 0.4)
$P(X \leq 4) = 0.0510$ (i.e. greater than the significance level of 0.05).
This means there is not enough evidence to reject the null hypothesis.
Hence the manager's statement is unlikely to be correct.

For the alternative hypothesis, the probability of him winning has increased as he has improved.

4 **Method using *p*-values**

$\mathbf{H_0}$: $p = 0.4$ $\mathbf{H_1}$: $p > 0.4$

Test statistic X is the number of games he wins.

X is B(8, 0.4)

$P(X \geq 6) = 1 - P(X \leq 5)$

$= 1 - 0.9502$

$= 0.0498$ (i.e. less than the significance level of 5%).

As 0.0498 < 0.05, there is evidence to reject the null hypothesis in favour of the alternative hypothesis. This means there is evidence that his game playing skills have improved.

Method using critical value

Look down the column in the table for the first probability that exceeds 0.95.
$P(X \leq 5) = 0.9502$ which means the critical value is one after it. So the critical value is 6 and the critical region is $X \geq 6$.
Hence the critical region is 6, 7 and 8.
As 6 games won is in the critical region the null hypothesis is rejected in favour of the alternative hypothesis so there is evidence that his game playing skills have improved.

5 (a) (i) Opportunity sampling as he has just used a convenient sample (i.e. a class of students he is teaching).

(ii) The sampling method can give unrepresentative results. A-level students are in the sixth form and are much more likely to be given a greater number of hours homework.

(b) (i) $\mathbf{H_0}$: $p = 0.5$
$\mathbf{H_1}$: $p < 0.5$

(ii) X is the number of students who got at least one hour's homework each day in the previous week.
X is B(10, 0.50)
$P(X \leq 2) = 0.0547$

Note that fewer than 4 means 3 or fewer students.

(iii) $P(X \leq 3) = 0.1719$

(c) Level of significance = 0.05
$P(X \leq 1) = 0.0107$. As this value is less than 0.05 the critical region is $X \leq 1$. So values in the critical region are 0 and 1.
The value $X = 2$ in the sample is not in the critical region so the null hypothesis is not rejected. This means the head teacher's belief is supported by the evidence, so he is probably correct.

BOOST Grade

It is not sufficient to say that the null hypothesis is accepted. You need to give the answer in the context of the question (i.e. if the head teacher's belief is correct or incorrect).

6 (a) Null hypothesis $\mathbf{H_0}$:$p = 0.35$
Alternative hypothesis $\mathbf{H_1}$:$p < 0.35$

(b) X is B(20, 0.35)

(c) (i) Assuming the null hypothesis is true, we use B(20, 0.35).
As there is a '<' sign in the alternative hypothesis, we use the lower tail of the probability distribution.
Using the binomial distribution function table and looking down the column for the first probability that exceeds 0.05 we find $P(X \leq 4) = 0.1182$. The critical value, a, is the value before this, so the critical value is 3.

(ii) The critical region is $X \leq 3$ (i.e. $X = 0, 1, 2, 3$).

(d) As $X = 6$, this is outside the critical region (i.e. in the acceptance region). Hence the null hypothesis is not rejected and it suggests Alex has not overestimated her support.

7 In this answer we will use the p-value approach but you could equally determine the critical value and critical region and see if 10 lies in it.
Null hypothesis $\mathbf{H}_0 : p = 0.45$
Alternative hypothesis $\mathbf{H}_1 : p > 0.45$
X is B(15, 0.45)
$P(X \geq 10) = 1 - P(X \leq 9)$
$= 1 - 0.9231$
$= 0.0769$
Now $0.0769 > 0.05$ (i.e. the significance level)
Hence the null hypothesis is not rejected and the player is probably correct in his assertion and the manager is probably wrong.

This is a one-tailed test in the direction of the upper tail. Notice the '>' sign in the alternative hypothesis.

Remember that the tables give us the probabilities of values at most equal to a certain value of X.

8 (a) Null hypothesis $\mathbf{H}_0 : p = 0.15$
Alternative hypothesis $\mathbf{H}_1 : p < 0.15$
(b) X is B(50, 0.15)
$P(X \leq 3) = 0.0460$
Now $0.0460 < 0.05$ so we reject $\mathbf{H}_0$, which suggests the production manager is probably incorrect.

The '<' sign tells us we are looking at the lower tail of the probability distribution.

In this question we have used the p-value method rather than the method that involves finding the critical value and critical region.

9 p is the probability of Hope selling a car to a customer who has booked a test drive.
The null hypothesis is based on Hope's belief, so $\mathbf{H}_0: p = 0.43$.
The alternative hypothesis is the manager's claim that this probability is higher than this, so $\mathbf{H}_1: p > 0.43$ (note the > sign means we need to look at the upper tail).
X, the test statistic, is the number of cars sold after a test drive
X is B(30, 0.43).
The level of significance, $\alpha = 0.01$.

The p-value is the probability that under the null hypothesis, the value of X is at least as extreme as the observed value.
Hence $P(X \geq 18)$ needs to be found using tables or a calculator.
Now to use the tables we need to find $1 - P(X \leq 17) = 1 - 0.9544 = 0.0456$
Now $0.0456 > 0.01$ so the result is not significant.
There is insufficient evidence at the 1% level of significance to reject the null hypothesis in favour of the manager's belief.

Note that as $p = 0.43$ we cannot use binomial tables as there are no values for $p = 0.43$. Instead we must use a calculator.

This probability is called the p-value. Make sure you don't get this value mixed up with the p used in the binomial distribution.

10 (a) Two reasons from the following:
- There is only success or failure (i.e. faulty or not faulty).
- There are trials with a constant probability of success.
- There are a fixed number of trials (i.e. $n = 50$).
- The trials are independent.

(b) X can be modelled by B(50, 0.25).
This is a two-tailed test, so the critical region consists of a region at the end of each tail.
The significance level is divided by two to give the probability at each tail (i.e. 0.025).
Using the tables:
Considering the tail to the left and finding the first value that exceeds 0.025, we have $P(X \leq 7) = 0.0453$ and the critical value is the value before it so $X = 6$ is the critical value and the critical region for this tail is $X \leq 6$.
Considering the tail to the right and using the tables to find a value

of X that is at least $1 - 0.025 = 0.975$. From the tables $P(X \leq 19) = 0.9861$. This means we pick the value after this, so $X = 20$ is the critical value and $X \geq 20$ is the critical region.

Hence the critical values are 6 and 20 and the critical regions are $X \leq 6$ and $X \geq 20$.

Section B Mechanics

Topic 6

No questions

Topic 7

1 (a) $u = 0\ \text{m s}^{-1}$, $a = 0.9\ \text{m s}^{-2}$, $t = 10\ \text{s}$, $v = ?$

Using
$$\begin{aligned} v &= u + at \\ &= 0 + 0.9 \times 10 \\ &= 9\ \text{m s}^{-1} \end{aligned}$$

(b) Using
$$\begin{aligned} s &= \tfrac{1}{2}(u + v)t \\ &= \tfrac{1}{2}(0 + 9)10 \\ &= 45\ \text{m} \end{aligned}$$

2 (a) Taking the upward velocity as positive, we have

$u = 20\ \text{m s}^{-1}$, $v = 0\ \text{m s}^{-1}$, $g = -9.8\ \text{m s}^{-2}$

Using $v^2 = u^2 + 2as$ gives $0 = 20^2 + 2 \times (-9.8) \times s$

Solving for s, gives $s = 20.4\ \text{m}$

(b) Using $s = ut + \frac{1}{2}at^2$

The displacement, s, is zero when the stone returns to its point of projection so

$$0 = 20t + \tfrac{1}{2} \times (-9.8) \times t^2$$
$$0 = 20t - 4.9t^2 = t(20 - 4.9t)$$
$$t = 0 \text{ or } 4.1\ \text{s}$$

$t = 0\ \text{s}$ is ignored as a possible time

Hence time = 4.1 s

3 (a) Taking the downward direction as positive.

$u = 0.8\ \text{m s}^{-1}$, $t = 3.5\ \text{s}$, $a = g = 9.8\ \text{m s}^{-2}$, $v = ?$

Using $v = u + at$ we have $v = 0.8 + 9.8 \times 3.5 = 35.1\ \text{m s}^{-1}$

(b) Using $s = ut + \frac{1}{2}at^2 = 0.8 \times 3.5 + \frac{1}{2} \times 9.8 \times 3.5^2 = 62.8\ \text{m}$

4 (a) Taking upwards as the positive direction, we have

$u = 10\ \text{m s}^{-1}$, $a = g = -9.8\ \text{m s}^{-2}$, $v = 0\ \text{m s}^{-1}$

Using $v = u + at$ gives

$$0 = 10 - 9.8t$$

Hence, $t = 1.02\ \text{s}$

(b) Using $v^2 = u^2 + 2as$ gives

$$0 = 10^2 + 2 \times (-9.8)s$$

Hence $s = 5.1\ \text{m}$

At its greatest height, the velocity will be zero.

5 (a) $v = \frac{dr}{dt} = 36t^2$

(b) $a = \frac{dv}{dt} = 72t$

When $t = 2$ s, $a = 72t = 72 \times 2 = 144$ m s^{-2}

6 (a) $a = \frac{dv}{dt} = 1.92t^2 - 0.72t$

(b) $r = \int v\,dt$

$= \int(0.64t^3 - 0.36t^2)dt$

$= \frac{0.64t^4}{4} - \frac{0.36t^3}{3} + c$

$= 0.16t^4 - 0.12t^3 + c$

When $t = 0$, $r = 0$ and substituting these values gives

$0 = 0 - 0 + c$

Hence $c = 0$.

So $r = 0.16t^4 - 0.12t^3$

When $t = 10$ s, $r = 0.16(10)^4 - 0.12(10)^3 = 1480$ m

7 (a) $v = \int a\,dt = \int(3 - 0.1t)dt = 3t - \frac{0.1t^2}{2} + c$

When $t = 0$ s, $v = 0$ m s^{-1}

Hence $0 = 3(0) - \frac{0.1(0)^2}{2} + c$, giving $c = 0$

$v = 3t - \frac{0.1t^2}{2}$

(b) $v = 3(10) - \frac{0.1(10)^2}{2} = 25$ m s^{-1}

(c) When $t = 30$ s, $a = 3 - 0.1t = 3 - 0.1 \times 30 = 0$ m s^{-2}

The lorry is speeding up before $t = 30$ and slowing down after $t = 30$, and is instantaneously at rest at $t = 30$.

(d) $r = \int v\,dt$

$= \int(3t - 0.05t^2)dt$

$= \frac{3t^2}{2} - \frac{0.05t^3}{3} + c$

When $t = 0$, $r = 0$ so $0 = \frac{3(0)^2}{2} - \frac{0.05(0)^3}{3} + c$, giving $c = 0$

Hence $r = \frac{3t^2}{2} - \frac{0.05t^3}{3}$

When $t = 30$ s, $r = \frac{3(30)^2}{2} - \frac{0.05(30)^3}{3} = 1350 - 450 = 900$ m

8 $r = \int v\,dt$

$= \int(6t^2 + 4)dt$

$= \frac{6t^3}{3} + 4t + c$

$= 2t^3 + 4t + c$

When $t = 0$, $r = 0$ so we have $0 = 2(0)^3 + 4(0) + c$ and solving gives $c = 0$

Hence $r = 2t^3 + 4t$

When $t = 2$ s, $s = 2(2)^3 + 4(2) = 24$ m

When $t = 5$ s, $s = 2(5)^3 + 4(5) = 270$ m

Distance travelled between the times $t = 2$ s and $t = 5$ s is $270 - 24 = 246$ m

Rather than use this method you could have used definite integration using the limits of 5 and 2 to find the distance travelled between these two times. In the exam you are free to choose your method.

The particle is at the origin at $t = 0$ so we know $r = 0$ as the origin is the point from which the distance is measured.

9 (a) (i) $a = \frac{dv}{dt}$
$= 12t - 2$
(ii) When $t = 1$, $a = 12(1) - 2 = 10\text{ m s}^{-2}$
(b) $r = \int v\,dt$
$= \int (6t^2 - 2t + 8)\,dt$
$= \frac{6t^3}{3} - \frac{2t^2}{2} + 8t + c$
$= 2t^3 - t^2 + 8t + c$
When $t = 0$, $r = 0$ so $0 = 2(0)^3 - (0)^2 + 8(0) + c$
Solving gives $c = 0$
Hence, the expression for the displacement is
$r = 2t^3 - t^2 + 8t$

10 (a) $153\text{ km h}^{-1} = \frac{153 \times 1000}{3600} = 42.5\text{ m s}^{-1}$ and $0\text{ km h}^{-1} = 0\text{ m s}^{-2}$
$v = u + at$
$a = \frac{v - u}{t} = \frac{0 - 42.5}{2} = -21.25\text{ m s}^{-1}$
Deceleration = 21.25 m s^{-1}
(b) Distance travelled whilst coming to rest, $s = ut + \frac{1}{2}at^2$
$s = 42.5 \times 2 + \frac{1}{2} \times (-21.25) \times 2^2 = 42.5\text{ m}$
Fraction of deck used = $\frac{42.5}{300} = \frac{17}{120}$

Remember to always simplify fractions.

11 (a) Distance travelled = area of trapezium = $\frac{1}{2}(30 + 15) \times 6 = 135\text{ m}$
(b) Distance travelled between $t = 30$ s and $t = 55$ s = area of triangle
$= \frac{1}{2} \times 25 \times 6 = 75\text{ m}$
Total distance travelled = 135 + 75 = 210 m
(c) Displacement in the positive direction = 135 m and displacement in the negative direction = 75 m
Overall displacement from its original position = 135 − 75 = 60 m
(d) Average speed = $\frac{\text{total distance travelled}}{\text{time taken}} = \frac{210}{55} = 3.8\text{ m s}^{-1}$ (2 s.f.)

12 (a) First list the letters and their values when they are known.
$u = 2\text{ m s}^{-1}$, $a = 9.8\text{ m s}^{-2}$, $t = 3$ s, $s = ?$
Using $s = ut + \frac{1}{2}at^2$ we have
$s = 2 \times 3 + \frac{1}{2} \times 9.8 \times 3^2 = 50.1\text{ m}$

Here the downward direction is taken as positive.

(b) $v = u + at$
$= 2 + 9.8 \times 3$
$= 31.4\text{ m s}^{-1}$
(c) The stone is modelled as a particle (i.e. you can ignore its dimensions). There is no air resistance.

13 (a) The constant velocity of the speeding car is represented by a horizontal line and the accelerating police car is represented by a line at an angle from $t = 2$ s.

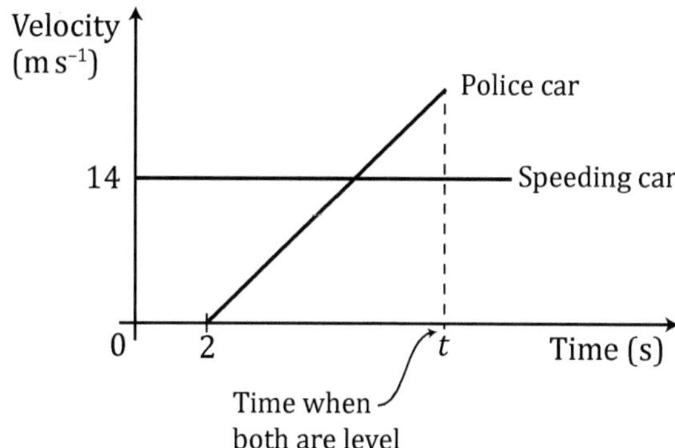

Note the point of intersection of these lines represents when the velocities are the same and not where they draw level.
If we let the time they draw level = t s, then the area under the rectangle for the speeding car will be equal to the area of the triangle under the line representing the motion of the police car.

(b) Distance travelled by speeding car = area of rectangle = $14t$
To work out the distance travelled by the police car we need to work out the area of the triangle.
The base of the triangle will have a length of $t - 2$. In order to work out an expression for the distance travelled by the police car, we need to work out the height of the triangle. This is the velocity reached by the car after travelling $t - 2$ s from rest with an acceleration of 4 m s^{-2}.
Using $v = u + at$

$$v = 0 + 4(t - 2)$$
$$v = 4(t - 2)$$

Now the area of the triangle

$$= \frac{1}{2} \times \text{base} \times \text{height} = \frac{1}{2}(t - 2)4(t - 2) = 2(t - 2)^2$$

When they draw level, their distances are the same so you can equate these areas.

$$2(t - 2)^2 = 14t$$
$$(t - 2)^2 = 7t$$
$$t^2 - 4t + 4 = 7t$$
$$t^2 - 11t + 4 = 0$$

Note that this quadratic equation will not factorise so the formula is used.

$$\frac{-b \pm \sqrt{b^2 - 4ac}}{2a} = \frac{-(-11) \pm \sqrt{(-11)^2 - 4(1)(4)}}{2(1)} = \frac{11 \pm \sqrt{105}}{2} = 10.6 \text{ s or } 0.377 \text{ s}$$

Now 0.337 s is impossible as this is before the police car sets off.
Hence time when both are level = 11 s (2 s.f.)

(c) As both the speeding and police cars travel the same distance when they draw level
Distance travelled by police car = $14t = 14 \times 10.6 = 148.4$ m = 150 m (2 s.f.)

14 (a) $v = t^3 - 2t^2 + t$ so when $v = 0$, $t^3 - 2t^2 + t = 0$

$$t(t^2 - 2t + 1) = 0$$
$$t(t - 1)(t - 1) = 0$$

Solving gives $t = 0$ s or 1 s

> Remember that to differentiate you multiply by the index and then reduce the index by 1.

> You could use the alternative method where you integrate and find the constant of integration. You then put c back to give the equation which is then used to find the distances travelled at the two times. These are subtracted to find the distance travelled between the two times.

(b) $a = \frac{dv}{dt}$

$= 3t^2 - 4t + 1$

(c) $s = \int_0^1 v\,dt$

$= \int_0^1 (t^3 - 2t^2 + t)dt$

$= \left[\frac{t^4}{4} - \frac{2t^3}{3} + \frac{t^2}{2}\right]_0^1$

$= \left[\frac{1}{4} - \frac{2}{3} + \frac{1}{2}\right] - 0$

$= \frac{1}{12}$ m

15 Need to first find the value of t which corresponds to the stationary points.

$$\frac{dr}{dt} = 6t - 12t^2$$

r has its stationary values (i.e. maximum and minimum values) when $\frac{dr}{dt} = 0$

Hence $6t - 12t^2 = 0$

$6t(1 - 2t) = 0$

Solving gives $t = 0$ or $t = \frac{1}{2}$. $t = 0$ is ignored as this will be at the surface and will therefore be when the ball is at its minimum depth.

Hence the maximum value of r is when $t = 0.5$ s, so, as $r = 3t^2 - 4t^3$, we have

$$r = 3(0.5)^2 - 4(0.5)^3 = 0.25 \text{ m}$$

Max depth = 0.25 m

Topic 8

1

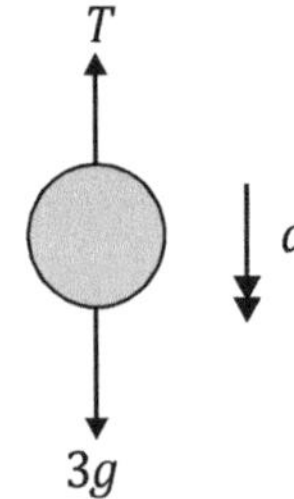

> We can ignore the weight of a light string but not the weight of a heavy chain.

> The weight of the particle acts vertically down and is given by $W = mg$. As the mass here is 3 kg, $W = 3g$.

There is no resultant force in the horizontal direction.
Applying Newton's second law in the vertical direction gives

$$ma = 3g - T$$

$$3 \times 2 = (3 \times 9.8) - T$$

giving $T = 23.4$ N

2 (a) Applying Newton's second law of motion to the 1.5 kg mass, we obtain

$$ma = T - 1.5g$$

$$1.5a = T - 1.5g \qquad (1)$$

Applying Newton's second law of motion to the 2 kg mass, we obtain

$$ma = 2g - T$$

$$2a = 2g - T \qquad (2)$$

Solving equations (1) and (2) simultaneously, we obtain

$$T = 16.8\text{ N}$$

(b) Solving equations (1) and (2) simultaneously, we obtain

$$a = 1.4\text{ m s}^{-2}$$

(c) When both particles are 1 m apart, the 1.5 kg mass would have risen by 0.5 m and the 2 kg mass would have fallen by 0.5 m.

For the 2 kg mass, $u = 0$, $a = 1.4$, $s = 0.5$, $v = ?$

Using $\quad v^2 = u^2 + 2as,$

$$v^2 = 0^2 + 2 \times 1.4 \times 0.5$$

Hence $\quad v = 1.18\text{ m s}^{-1}$

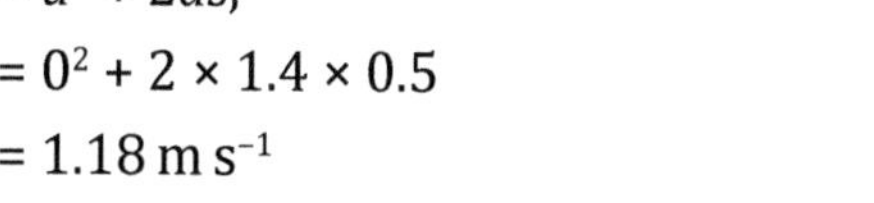

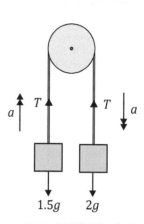

T, T, a, a, 1.5g, 2g

Use the equations of motion to find the final velocity, v.

Remember that the equations of motion are not in the formula booklet and must therefore be memorised.

3 (a) Applying Newton's second law to particle A, we obtain

$$5a = T - 5g$$

$$10 = T - (5 \times 9.8)$$

$$T = 59\text{ N}$$

(b) Applying Newton's second law to particle B, we obtain

$$m \times 2 = mg - T$$

$$2m = mg - T$$

$$2m = 9.8m - 59$$

$$m = 7.6\text{ kg}$$

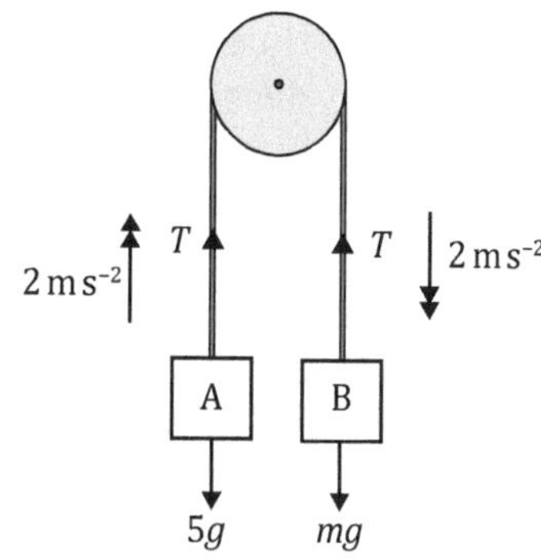

4 (a) Applying Newton's 2nd law of motion to mass P, we obtain

$$ma = mg - T$$

$$5a = 5g - T \qquad (1)$$

Applying Newton's 2nd law of motion to mass Q, we obtain

$$ma = T - 2g$$

$$2a = T - 2g \qquad (2)$$

Solving equations (1) and (2) simultaneously we obtain:

$$a = 4.2\text{ m s}^{-2} \text{ and } T = 28\text{ N}$$

(b) The light string allows the assumption to be made that the tension in the string remains constant throughout the string.

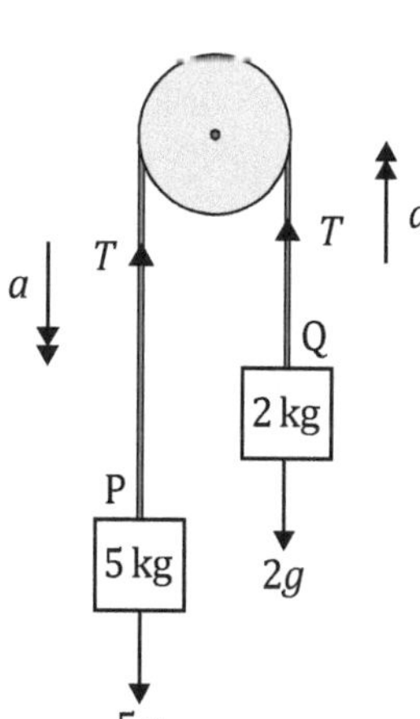

Mass P will accelerate downwards as its weight is larger. Both masses will accelerate with the same magnitude of acceleration.

Note that this is not the usual pulley-type question as you are not asked to find the acceleration or the tension in the string. You do need to find the velocity with which the 3 kg mass hits the ground so you need to find the acceleration and then use the equations of motion.

This will be the constant acceleration of the 3kg mass as it falls from rest.

5 First draw a diagram and mark on the forces, accelerations and distance.

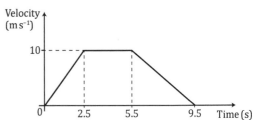

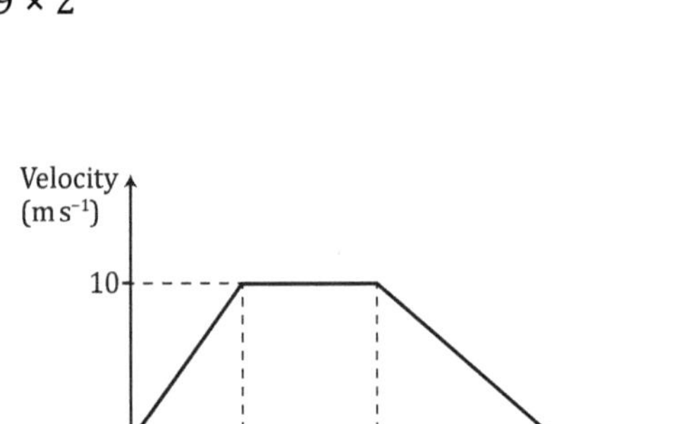

Applying Newton's 2nd law to the 3 kg mass, we obtain

$$3a = 3g - T \qquad (1)$$

Applying Newton's 2nd law to the 1 kg mass, we obtain

$$a = T - g \qquad (2)$$

Adding equations (1) and (2) we obtain

$$4a = 2g$$

$$a = \tfrac{1}{2}g$$

$$a = 4.9\ \text{m s}^{-2}$$

$u = 0\ \text{m s}^{-1}$, $a = 4.9\ \text{m s}^{-2}$, $s = 2$ m, $v = ?$

Using $v^2 = u^2 + 2as$

$$v^2 = 0^2 + 2 \times 4.9 \times 2$$

$$v = 4.4\ \text{m s}^{-1}$$

(b) There is no air resistance.

6 (a) Note that you can find the times for the acceleration and deceleration by dividing 10 m s^{-1} by the acceleration/deceleration. This enables you to mark some values on the time axis.

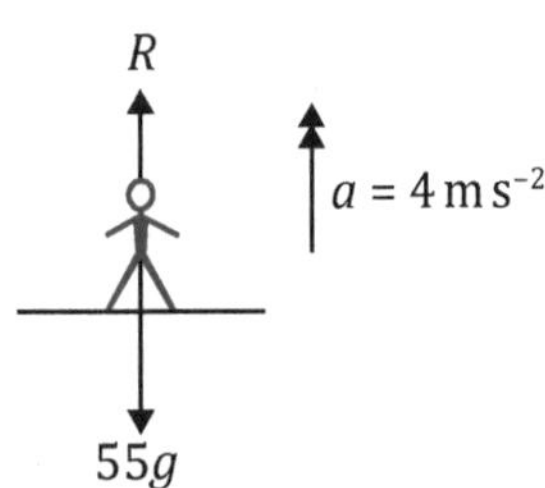

(b) Total distance travelled = area under the velocity–time graph

= area of the trapezium

$= \tfrac{1}{2}(9.5 + 3)10$

= 62.5 m

(c) The maximum reaction occurs when the lift accelerates.

Applying Newton's 2nd law of motion to the person in the lift we obtain

$$55a = R - 55g$$

$$55 \times 4 = R - 55 \times 9.8$$

$$R = 759\ \text{N}$$

Note that when the lift is travelling at constant velocity the reaction is equal to the weight so this situation need not be considered.

7 (a) Applying Newton's 2nd law to the lift, we obtain

$$ma = T - 500g$$

$$500a = 6000 - 500g$$

$$a = 2.2\ \text{m s}^{-1}$$

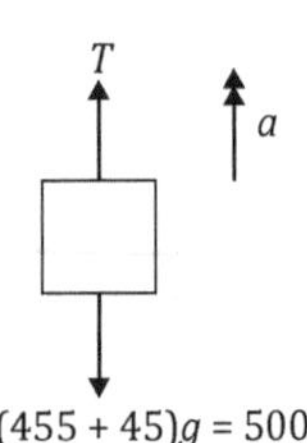

(b) Applying Newton's 2nd law to the man in the lift, we have

$$45 \times 2.2 = R - 45 \times 9.8$$

$$R = 540\ \text{N}$$

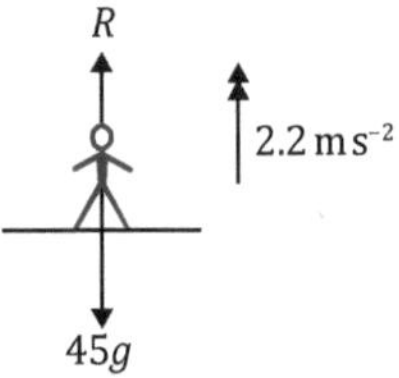

8 First draw a diagram even though one is not asked for. Mark on all the forces acting and the directions of the accelerations.

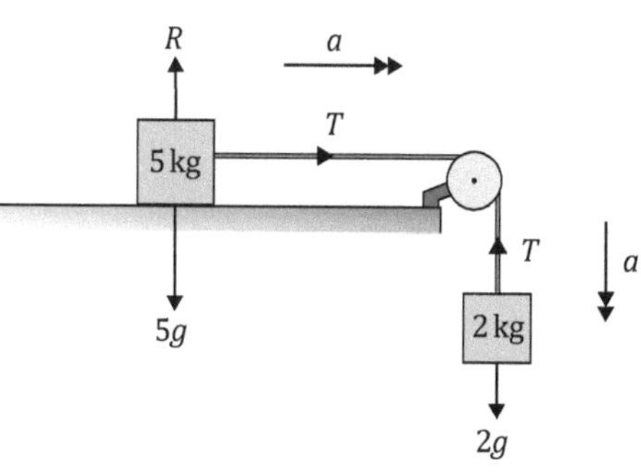

Applying Newton's 2nd law of motion to the 5 kg mass, we obtain

$$ma = T$$
$$5a = T \qquad (1)$$

Applying Newton's 2nd law of motion to the 2 kg mass, we obtain

$$ma = 2g - T$$
$$2a = 19.6 - T \qquad (2)$$

Adding equations (1) and (2), we obtain

$$7a = 19.6$$
$$a = 2.8 \text{ m s}^{-2}$$

Substituting $a = 2.8$ into equation (1) gives

$$T = 5 \times 2.8 = 14 \text{ N}$$

9 (a) A diagram is drawn showing the forces acting and the directions of the accelerations.

Applying Newton's 2nd law of motion to P, we obtain

$$ma = T$$
$$8a = T \qquad (1)$$

Applying Newton's 2nd law of motion to mass Q, we obtain

$$ma = 5g - T$$
$$5a = 49 - T \qquad (2)$$

Adding equations (1) and (2), we obtain

$$13a = 49$$
$$a = 3.8 \text{ ms}^{-2}$$

Substituting $a = 3.8$ into equation (1) gives

$$T = 8 \times 3.8 = 30.4 \text{ N}$$

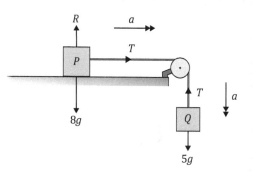

(b) Applying Newton's 2nd law of motion to P, we obtain

$8a = T$ so $T = 8 \times 5 = 40$ N

Applying Newton's 2nd law of motion to mass Q, we obtain

$(5 + M)a = (M + 5)g - T$

Substituting in $a = 5$ and $T = 40$ gives

$(5 + M)5 = (M + 5)9.8 - 40$

$25 + 5M = 9.8M + 49 - 40$

$16 = 4.8M$

$M = 3.3$ kg

Topic 9

1 (a) Resultant of the two forces = $\mathbf{L} + \mathbf{M} = (5\mathbf{i} + 9\mathbf{j}) + (2\mathbf{i} + 15\mathbf{j}) = 7\mathbf{i} + 24\mathbf{j}$

Using Newton's 2nd law: $\mathbf{F} = m\mathbf{a}$

$$7\mathbf{i} + 24\mathbf{j} = 5\mathbf{a}$$
$$\mathbf{a} = \tfrac{7}{5}\mathbf{i} + \tfrac{24}{5}\mathbf{j}$$

(b) $|\mathbf{a}| = \sqrt{\left(\frac{7}{5}\right)^2 + \left(\frac{24}{5}\right)^2} = \sqrt{25} = 5 \text{ m s}^{-2}$

(c)

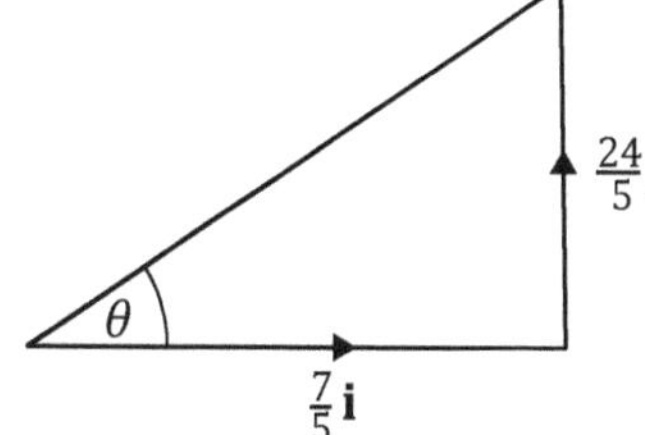

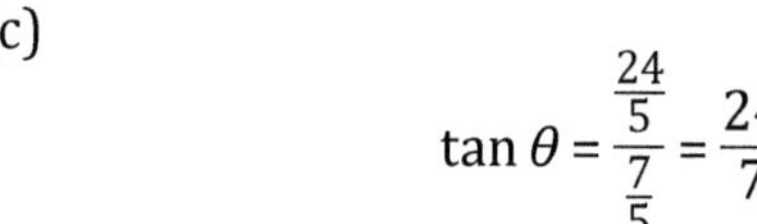

$$\tan\theta = \frac{\frac{24}{5}}{\frac{7}{5}} = \frac{24}{7}$$

$$\theta = \tan^{-1}\left(\frac{24}{7}\right) = 74\,° \text{ (nearest degree)}$$

Test yourself answers

2 Let $\mathbf{F} = a\mathbf{i} + b\mathbf{j}$

$$(2\mathbf{i} - 3\mathbf{j}) + (\mathbf{i} + 6\mathbf{j}) + (-4\mathbf{i} - 4\mathbf{j}) + a\mathbf{i} + b\mathbf{j} = 7\mathbf{i} + 2\mathbf{j}$$
$$-\mathbf{i} - \mathbf{j} + a\mathbf{i} + b\mathbf{j} = 7\mathbf{i} + 2\mathbf{j}$$
$$a - 1 = 7$$
$$a = 8$$
$$b - 1 = 2$$
$$b = 3$$
$$\mathbf{F} = 8\mathbf{i} + 3\mathbf{j}$$

3 Displacement $= (3\mathbf{i} - 4\mathbf{j}) + (2\mathbf{i} + \mathbf{j}) + (-4\mathbf{i} + 6\mathbf{j}) + (5\mathbf{i} - \mathbf{j}) + (5\mathbf{i} + 3\mathbf{j}) = 11\mathbf{i} + 5\mathbf{j}$

4 Resultant force $= (4\mathbf{i} + \mathbf{j}) + (6\mathbf{i} - 6\mathbf{j}) = 10\mathbf{i} - 5\mathbf{j}$

$$\text{Force} = m \times \mathbf{a}$$
$$10\mathbf{i} - 5\mathbf{j} = 10 \times \mathbf{a}$$
$$\mathbf{a} = \frac{10\mathbf{i} - 5\mathbf{j}}{10} = \mathbf{i} - 0.5\mathbf{j}$$

5 (a) Resultant force $= (3\mathbf{i} + 6\mathbf{j}) + (-2\mathbf{i} - \mathbf{j}) + (2\mathbf{i} - \mathbf{j}) = (3\mathbf{i} + 4\mathbf{j})$N

(b) Magnitude of resultant force $= \sqrt{3^2 + 4^2} = \sqrt{25} = 5$ N

(c)

$$\tan\theta = \frac{4}{3} \quad \text{so } \theta = \tan^{-1}\left(\frac{4}{3}\right)$$
$$\theta = 53.1°$$

Bearing $= 90 - 53.1 = 36.9° = 037°$

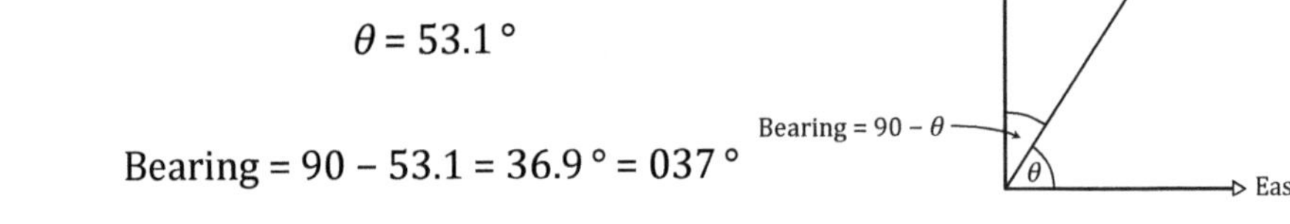

Bearings are the angle measured in a clockwise direction from due north. They are usually given as 3 figures to the nearest degree.

6 (a) Resultant force $= (\mathbf{i} + 3\mathbf{j}) + (2\mathbf{i} - 5\mathbf{j}) + (4\mathbf{i} + 6\mathbf{j}) = 7\mathbf{i} + 4\mathbf{j}$

(b)
$$\mathbf{F} = m \times \mathbf{a}$$
$$7\mathbf{i} + 4\mathbf{j} = 2\mathbf{a}$$
$$\mathbf{a} = 3.5\mathbf{i} + 2\mathbf{j}$$

(c) $|\mathbf{a}| = \sqrt{3.5^2 + 2^2} = 4.03 = 4\text{ m s}^{-2}$

$$\theta = \tan^{-1}\left(\frac{2}{3.5}\right) = 29.7° \text{ (nearest 0.1°)}$$

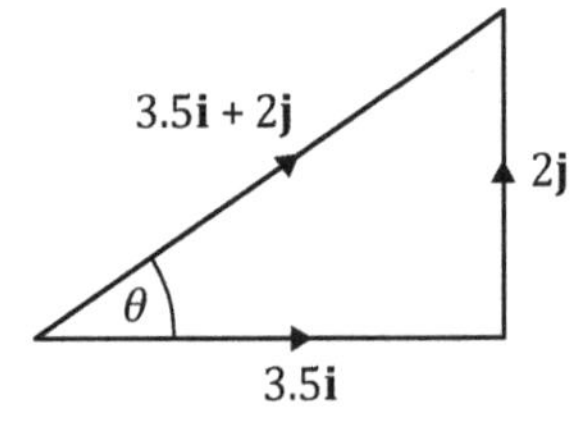

7 (a) $\overrightarrow{XZ} = (-2\mathbf{i} + 13\mathbf{j}) + (7\mathbf{i} - \mathbf{j}) = 5\mathbf{i} + 12\mathbf{j}$ km

(b) $|\overrightarrow{XZ}| = \sqrt{5^2 + 12^2} = 13$ km

(c) $\overrightarrow{XY} = -2\mathbf{i} + 13\mathbf{j}$ so $|\overrightarrow{XY}| = \sqrt{(-2)^2 + 13^2} = \sqrt{173}$

$\overrightarrow{YZ} = 7\mathbf{i} - \mathbf{j}$ so $|\overrightarrow{YZ}| = \sqrt{7^2 + (-1)^2} = \sqrt{50}$

Total distance from X to Z $= \sqrt{173} + \sqrt{50} = 20.2$ km (3 s.f.)

8 Using trigonometry:

$$\frac{a}{15} = \cos 60°$$
$$a = 15\cos 60° = 15 \times \frac{1}{2} = 7.5$$
$$\frac{b}{15} = \sin 60°$$
$$b = 15\sin 60° = 15 \times \frac{\sqrt{3}}{2} = 7.5\sqrt{3}$$

Vector is $\mathbf{v} = 7.5\mathbf{i} + 7.5\sqrt{3}\mathbf{j}$

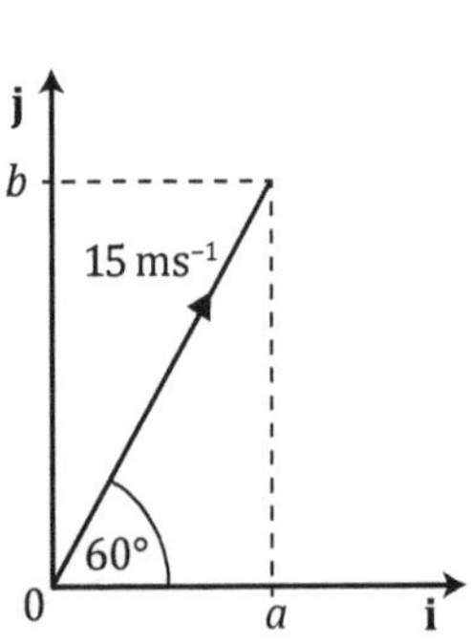

Notice the question asks for exact values for *a* and *b*. This means that you must keep the square root and not work it out as a decimal.